A GATEWAY TO
ABSTRACT
MATHEMATICS

A GATEWAY TO
ABSTRACT
MATHEMATICS

BY

E. A. MAXWELL, Ph.D.
Fellow of Queens' College, Cambridge

CAMBRIDGE
AT THE UNIVERSITY PRESS
1965

CAMBRIDGE UNIVERSITY PRESS
Cambridge, New York, Melbourne, Madrid, Cape Town, Singapore, São Paulo, Delhi

Cambridge University Press
The Edinburgh Building, Cambridge CB2 8RU, UK

Published in the United States of America by Cambridge University Press, New York

www.cambridge.org
Information on this title: www.cambridge.org/9780521057011

© Cambridge University Press 1965

First published 1965
This digitally printed version 2008

A catalogue record for this publication is available from the British Library

Library of Congress Catalogue Card Number: 65–19155

ISBN 978-0-521-05701-1 hardback
ISBN 978-0-521-09028-5 paperback

CONTENTS

CONTENTS

PREFACE

The subject-matter of most of the topics developed in this book is believed to be essentially new, though, of course, the ideas have been both foreshadowed and overtaken by many other writers. The aim is to provide material, familiar in substance but unfamiliar in treatment, that may catch the interest of pupils (and, dare I say, of teachers?) as they cross into the somewhat puzzling world of abstract mathematics. I have often felt that the present plunge into abstraction is too sudden and that there is a need for more elementary work to make the immersion less exhausting.

To carry the subject forward from this stage will be the work of others; the hope is that this book may help to ease the start.

I am very grateful to the staff of the Cambridge University Press for their skill and care, and for the friendly relations that I have always enjoyed with them.

E. A. M.

Cambridge
March, 1965

INTRODUCTION

Few of us pause for long to think about the rules that we apply automatically when performing mathematical calculations, say in arithmetic or algebra. Thus we 'factorise'

$$ab + ac = a(b + c)$$

or 'multiply out'

$$(2a + 7b)x = 2ax + 7bx$$

by laws which are so familiar—at any rate by the stage implied by the reading of this book—that no justification seems necessary.

There is no intention here of codifying these laws systematically, except perhaps by inference; the aim is rather to exhibit *alternative* systems of rules for calculation, after which the basic laws themselves may appear both clear and natural.

We propose, then, a process of abstraction, akin to the fundamental ideas on which much so-called 'modern' mathematics is based. The important point, though, is that the material itself from which the abstractions are to spring will be selected from topics likely to be familiar to everyone who has studied just enough mathematics to reach 'O'-level. The great difficulty in introducing 'modern' mathematics seems to be the production of a background of experience from which it can start; and the avowed aim of this book is the attempt to foster just such experience.

CHAPTER I

DIGITAL ARITHMETIC

¶1. The ordinary facts of elementary arithmetic are well known. It will be no trouble to most people who read this book to say that

$$9 + 7 = 16,$$
$$28 + 35 = 63,$$
$$9 \times 7 = 63,$$

or even $\qquad 17 \times 12 = 204.$

The rules are familiar and 'counting' is, essentially, easy.

What we propose to do in this chapter is to accept this ordinary arithmetic, but to modify it in one way. The result of this modification will be to make *computation* easier but, in return, to make *thought* harder for the beginner because of the element of abstraction that now comes in.

The new process will be called *digital arithmetic*. The reason for this name is that the operations of 'addition' and 'multiplication' with which we shall be concerned are to be the normal operations familiar in elementary arithmetic, *save that for the answers at each step only the units digit is to be retained.*

For example, whereas normally

$$8 + 7 = 15,$$
$$9 + 3 = 12,$$

we shall simply take

$$8 + 7 = 5,$$
$$9 + 3 = 2;$$

and whereas normally
$$8 \times 6 = 48,$$
$$7 \times 3 = 21,$$
we shall simply take
$$8 \times 6 = 8,$$
$$7 \times 3 = 1.$$

Note at once that we have radically altered the meanings of the two symbols $+$ and $\times$ and of the corresponding operations of addition and multiplication. In order to give a meaning to, say,

$$7 + 9,$$

we do *not* imagine seven apples and nine apples put into a box which thus contains sixteen apples; the physical interpretation has disappeared. We have said instead that $7 + 9$ is to mean 6 (a number reached by the process already described); and that is all about it. Our problem is not to justify the definition itself, for we are entitled to give any clearly-stated meaning that we wish to our words and symbols, but to justify instead the possibility of *using* the definition consistently and with significance. We therefore begin by examining the consistency of the addition.

¶ 2. Digital Addition

The laws of elementary arithmetic require us to add any collection of numbers without ambiguity. Suppose, for example, that we are given the three numbers

$$9, 7, 18.$$

We should say at once that the sum is 34 and should probably mean by that statement that

$$9 + 7 + 18 = 34.$$

On further reflection, we should probably agree also that the statement equally means any of the following:

$$9 + 18 + 7 = 34,$$
$$7 + 18 + 9 = 34,$$
$$7 + 9 + 18 = 34,$$
$$18 + 9 + 7 = 34,$$
$$18 + 7 + 9 = 34.$$

In other words, *the three numbers* 9, 7, 18, *taken in any order, always add up to the same number* 34. A corresponding result holds similarly for any other set of numbers.

DIGRESSION. It may be helpful to give now as a digression what is in fact the ultimately basic property of addition. The fundamental facts of elementary arithmetic (as popularly understood) are a whole collection of answers obtained successively by adding *two* numbers at a time. For instance, to add

$$3 + 7 + 11 + 19 + 5,$$

we should probably say something like this:

$$3 + 7 = 10,$$
$$10 + 11 = 21,$$
$$21 + 19 = 40,$$
$$40 + 5 = 45.$$

At any rate, the process would almost certainly involve the addition step by step of *two* numbers at any one time.

Reverting to the simpler case of three numbers only, say

$$3 + 5 + 9,$$

there thus appears a choice:
    (i) to group $3 + 5$ and then add 9, giving

$$8 + 9,$$

or   (ii) to take 3 and then add to it the group $5 + 9$, giving

$$3 + 14.$$

The point is, of course, that these two groupings give the same answer. In other words,

$$(3+5)+9 \;=\; 3+(5+9).$$

More generally, *if a, b, c are any three of the numbers of ordinary arithmetic, and if the symbol + has its ordinary meaning, then*

$$(a+b)+c \;=\; a+(b+c).$$

This law, governing the two ways in which the three numbers can be associated for addition two at a time (without changing order) is called the *associative law* for addition.

It will have been noticed that not only can the numbers be grouped in pairs according to the associative law, but also, as the preceding statement of that law implied, that the actual order in which the numbers are written is also immaterial. For example,

$$5+9 \;=\; 9+5;$$

and, more generally, for ordinary arithmetic

$$a+b \;=\; b+a.$$

Two numbers whose order can be changed in this way are said to *commute* and the general law

$$a+b \;=\; b+a$$

is called the *commutative law* for addition.

EXAMPLE

Check mentally that the two laws just enunciated ensure that the sum of the five numbers

$$3, \; 7, \; 12, \; 15, \; 19$$

is independent of the order in which they are selected.

Returning to *digital* addition, consider a sum such as

$$5+7+9+3+8.$$

The normal sum is 32 and so the digital sum is 2. The problem, however, is: do the two laws, the associative and the commutative, still hold when the arithmetic is digital? The answer is clearly *yes*; for what is true of the complete numbers must *a fortiori* be true of the units digits. Hence *it remains true in digital arithmetic that the associative law for addition*

$$(a+b)+c = a+(b+c)$$

and the commutative law for addition

$$a+b = b+a$$

retain their validity.

Consequently *all arithmetical manipulations dependent on them, and valid in ordinary arithmetic, are equally valid in digital arithmetic.*

EXAMPLE

Check mentally that these two laws ensure that the 'sum' of the numbers

2, 5, 8, 3, 7

is independent of the order in which they are selected.

¶ 3. DIGITAL MULTIPLICATION

The corresponding treatment of digital multiplication may be discussed more briefly. The analogous laws are:

the *associative law* for multiplication,

$$(a \times b) \times c = a \times (b \times c)$$

or, more briefly,

$$(ab)c = a(bc);$$

and the *commutative law* for multiplication,

$$a \times b = b \times a,$$

or
$$ab = ba.$$

Since these laws hold for ordinary multiplication, they necessarily hold also for the units digits in the products.

EXAMPLE

Check mentally that these two laws ensure that the 'product' of the numbers

$$3, 5, 8, 4, 2, 6$$

is independent of the order in which they are selected.

❰ 4. DIVISORS OF ZERO

Without over-emphasising the point for the present, it is of interest to demonstrate at once a feature wherein digital arithmetic differs radically from ordinary arithmetic.

In ordinary arithmetic, as is well known, a relation

$$ab = 0$$

cannot hold unless at least one of a, b is zero. This is, for example, the basis of the argument used to finish the solution of a quadratic equation once it has been reduced to the form, say,

$$(x - 1)(x - 2) = 0.$$

The ending is 'Either $x - 1 = 0$ or $x - 2 = 0$, and so the equation is solved when $x = 1$ and when $x = 2$.

Note that this argument is only possible when the right-hand side is zero. It is *not true* that the equation

$$(x - 1)(x - 2) = 3$$

can be completed by the argument 'Either $x - 1 = 3$ or $x - 2 = 3$, and so the equation is solved when $x = 4$ and when $x = 5$.'

EXAMPLES

1. Prove that, if c is not zero and if a, b are different, then there is no set of numbers a, b, c for which the equation

$$(x - a)(x - b) = c$$

can be completed by the argument: 'Either $x-a = c$ or $x-b = c$ and so the equation is solved when $x = a+c$ and when $x = b+c$.'

2. Prove that, on the other hand, the equation

$$(x-3)(x-4) = 2$$

is satisfied (as to *one* solution) by setting $x-3 = 2$, so that $x = 5$; and that the equation

$$(4-x)(x-1) = 2$$

is satisfied (as to *both* solutions) by setting $4-x = 2$ or $x-1 = 2$, so that $x = 2$ or 3.

Examine the way in which these are specially constructed 'freak' equations and invent similar abnormalities yourself.

Returning to the main problem, we register the fact that, in ordinary arithmetic and algebra, a product cannot be zero unless one of its factors is. With digital arithmetic, however, the case is very different, in virtue of the four products

$$5 \times 2 = 0, \quad 5 \times 4 = 0, \quad 5 \times 6 = 0, \quad 5 \times 8 = 0.$$

In other words, *in digital arithmetic it is possible for the product*

$$ab$$

to be zero although neither factor is. This can happen when either one of a, b is 5 while the other is an even integer.

DEFINITION. A number a is called a *divisor of zero* when another number b exists such that the product ab is zero although neither individual factor is.

An immediate, and startling, consequence of the existence of divisors of zero is the possibility of producing in digital arithmetic *a quadratic equation with four distinct roots*:

Take, for example, the equation

$$(x-1)(x-2) = 0.$$

It is, of course, satisfied as usual by the two solutions

$$x = 1, \quad x = 2.$$

But there *may* also be a solution given by

$$x - 1 = 5$$

or by

$$x - 2 = 5,$$

provided that the second factor is an even integer. When $x = 6$, the left-hand side is 5×4, or zero; when $x = 7$, the left-hand side is 6×5, or zero. Hence *the quadratic* equation

$$x^2 - 3x + 2 = 0,$$

or

$$(x-1)(x-2) = 0,$$

has the *four* roots 1, 2, 6, 7.

EXAMPLES

1. Find all the solutions of the equations

    (i) $x^2 - 4x + 3 = 0$,

    (ii) $x^2 - 5x + 4 = 0$,

    (iii) $x^2 - 5x + 6 = 0$,

    (iv) $x^2 - 6x + 8 = 0$.

2. Prove that the quadratic equation

$$5x^2 + 5x = 0$$

is satisfied by all integral values of x. Explain why it is not permissible to divide throughout by 5 without further examination.

⟨ 5. THE DISTRIBUTIVE LAW

In order to demonstrate quickly the unexpected properties of digital arithmetic we have, in fact, glossed over one or two theoretical points of detail that now require attention. The first of these is the *distributive law*

$$\begin{cases} a(b+c) = ab+ac, \\ (b+c)a = ba+ca, \end{cases}$$

which enables us to 'remove brackets'. The truth of the law follows, as in the previous cases, from the fact that the law, being true for ordinary arithmetic, must, in particular, be true for the units digits.

It was this law that enabled us to use a sequence of argument (now given in detail) such as

$$(x+1)(x+2)$$

$$= (x+1)(x)+(x+1)(2) \quad \text{distributive}$$

$$= x^2+x+x(2)+1.2 \quad \text{distributive}$$

$$= x^2+x+2x+2 \quad \text{commutative}$$

$$= x^2+(x+2x)+2 \quad \text{associative}$$

$$= x^2+3x+2.$$

From now on we shall normally use the associative, commutative and distributive laws without comment. In other words, the manipulation will 'look like' ordinary algebra.

⟦6. The Language of Sets

Before passing to the next point of detail, we ought perhaps to say a few words about sets. Not much is required at present, but the language is convenient and leads to precision.

The effect of the definition of digital arithmetic is to replace the infinite sequence of numbers

$$..., -3, -2, -1, 0, 1, 2, 3, ...$$

by the *ten integers*

$$0, 1, 2, 3, 4, 5, 6, 7, 8, 9.$$

(The fate of the negative numbers is the next thing to be considered.) By ordinary use of language, we may say that the 'set'

$$\{..., -3, -2, -1, 0, 1, 2, 3, ...\}$$

has been replaced by the 'set'

$$\{0, 1, 2, 3, 4, 5, 6, 7, 8, 9\}.$$

These ideas give us two things—a *name* and a *notation*. The word *set* is used to denote any collection of objects whatever, subject to the sole requirement of a *rule* to determine whether a given object is a member or not; for example, *the set of all plane triangles* is defined: any given equilateral triangle is a member, but no circle can be.

A set may be designated for reference by any convenient letter, usually capital, and its members may be exhibited by enumeration within braces { }. Thus the set with which we are dealing in digital arithmetic may be denoted by the letter D, where

$$D \equiv \{0, 1, 2, 3, 4, 5, 6, 7, 8, 9\}.$$

Membership of a set is denoted by the *symbol of inclusion* ϵ, and non-membership by the symbol $\notin$. Thus

$$0 \in D, \quad 7 \in D, \quad \tfrac{1}{2} \notin D, \quad \pi \notin D.$$

¶ 7. At this point it is convenient to review rapidly the reader's experience with the numbers of ordinary arithmetic and algebra. It began, almost certainly, with the set of positive integers

$$M \equiv \{1, 2, 3, \ldots\}$$

and was probably extended fairly quickly to the corresponding set augmented by zero

$$N \equiv \{0, 1, 2, 3, \ldots\}.$$

After that we may imagine that there followed the rational fractions

$$F \equiv \{\tfrac{1}{1}, \tfrac{1}{2}, \tfrac{1}{3}, \tfrac{2}{3}, \tfrac{1}{4}, \tfrac{3}{4}, \tfrac{1}{5}, \ldots\}.$$

A little later, negative numbers would be studied, giving the set of integers

$$Z \equiv \{\ldots, -3, -2, -1, 0, 1, 2, 3, \ldots\}$$

and the set of all rational numbers

$$Q \equiv \{\ldots, -\tfrac{1}{3}, -\tfrac{1}{2}, -\tfrac{1}{1}, 0, \tfrac{1}{1}, \tfrac{1}{2}, \tfrac{1}{3}, \ldots\}.$$

Finally there may have come the set R of all *real* numbers, rational or irrational—the latter being described quickly as the numbers expressible as unending, non-recurring decimals—and the set C of *complex* numbers of the form $a + ib$, where a and b are real numbers as just described and where i is the 'square root of -1', subject to the relation $i^2 = -1$.

Most of this was, of course, unconscious, and it is well that it should have been so. But one basic point does arise from this classification of our number-experience into sets, and that point may now be illustrated by reference to a series of quadratic equations taken from the framework of ordinary arithmetic:

Consider in turn the quadratic equations:

$$\text{(i) } x^2 - 3x + 2 = 0, \quad \text{(iv) } 2x^2 + 3x + 1 = 0,$$
$$\text{(ii) } x^2 - 3x = 0, \quad\quad \text{(v) } x^2 - 3x + 1 = 0,$$
$$\text{(iii) } x^2 + 3x + 2 = 0, \quad \text{(vi) } x^2 - 3x + 3 = 0.$$

To the naked eye they are very similar, and, indeed, have been selected to be so. The distinction between them, as we shall now see, lies in the sets from which the solutions are drawn. The solutions are:

$$\text{(i) } 1, 2; \quad\quad\quad \text{(iv) } -1, -\tfrac{1}{2};$$
$$\text{(ii) } 0, 3; \quad\quad\quad \text{(v) } \tfrac{1}{2}(3 + \sqrt{5}), \tfrac{1}{2}(3 - \sqrt{5});$$
$$\text{(iii) } -1, -2; \quad \text{(vi) } \tfrac{1}{2}(3 + i\sqrt{3}), \tfrac{1}{2}(3 - i\sqrt{3}).$$

These solutions lie in the sets:

$$\text{(i) } M, N, F, Z, Q, R, C; \quad \text{(iv) } Q, R, C;$$
$$\text{(ii) } N, Z, Q, R, C; \quad\quad\quad \text{(v) } R, C;$$
$$\text{(iii) } Z, Q, R, C; \quad\quad\quad\quad \text{(vi) } C.$$

In other words, if the problem were to find a solution within the set Q, then equations (i), (ii), (iii), (iv) would be soluble, but *equations* (v), (vi) *would have no solutions*; if a solution were required within Z, then equations (i), (ii), (iii) would be soluble, but (iv), (v), (vi) *would have no solutions*.

The possibility of solving an equation thus depends on *the set from which solutions may be drawn.*

¶ 8. PROPERTIES OF THE NUMBERS IN DIGITAL ARITHMETIC

We said a little earlier (p. 19) that one or two difficulties in digital arithmetic were ignored in the opening remarks. Since then, the laws of manipulation have been clarified, and we can now turn our attention to *subtraction* and '*negative numbers*'.

Whatever is done with digital arithmetic must be accomplished within the set

$$D \equiv \{0, 1, 2, 3, 4, 5, 6, 7, 8, 9\},$$

and this is true of any meaning to be given to the word 'subtraction', to which we now proceed.

First consider (relevantly) the basic properties of an element, if any, to be regarded as *zero* and (less relevantly for the moment) of any element to be regarded as *unity*.

(a) *Zero.* The essential requirement is that, for $z \in D$ to be a possible zero, then, if $a \in D$ is any element whatever in D,

$$a + z = z + a = a;$$

in other words, the addition of zero makes no difference. A quick check reveals that this property belongs to 0 and to 0 only—as was probably anticipated.

Note incidentally the multiplicative property already observed, that, if a is any element of D, then

$$0 \times a = a \times 0 = 0.$$

It is, however, perfectly possible, because of the divisors of zero, for a product $a \times b$ to be zero although neither factor a, b is so itself.

(b) *Unity.* The requirement is that, for $z \in D$ to be a

possible unity, then, if $a \in D$ is any element whatever in D,

$$a \times z = z \times a = a;$$

in other words, multiplication by unity makes no difference. Here, again, the unity is quickly obtained as the expected number 1.

NOTE. The reader may have felt some impatience at these checks on zero and unity; but digital arithmetic is not ordinary arithmetic, and strange things may pass unnoticed if we are not careful; see p. 37.

¶ 9. We move now to the consideration of *negative numbers*. The problem is: *given any $a \in D$, to determine whether there is a number $x \in D$ to which we might reasonably give the notation* $-a$.

The obvious requirement is that, if possible, x should be selected so that

$$x + a = 0.$$

By the very definition of digital arithmetic, the number x is obtained from the formula

$$x = 10 - a$$

of ordinary arithmetic. Thus we interpret

$$-1, \; -2, \; -3, \; -4, \; -5, \; -6, \; -7, \; -8, \; -9$$

to be the numbers

$$9, \, 8, \, 7, \, 6, \, 5, \, 4, \, 3, \, 2, \, 1$$

of the set D.

The process of subtraction, intuitively 'obvious', now follows. For example,

$$7 - 4 = 7 + (-4) = 7 + 6 = 3,$$
$$4 - 7 = 4 + (-7) = 4 + 3 = 7,$$

and so on. The manipulation of subtraction is, in fact, just what would be expected.

¶ 10. DIVISION

The interpretation of a process of division is more complicated and is perhaps entertaining rather than useful. We take the opportunity to introduce two fresh symbols:

(i) The symbol

$$\forall$$

is used to mean 'for all', as in the sentence

$$\forall x, \ (-x) \text{ can be defined.}$$

(ii) The symbol

$$\exists$$

is used to mean 'there exist(s)', as in the sentence

$$\forall x, \ \exists y \text{ such that } x + y = 0;$$

or, in even more precise form,

$$\forall x \in D, \quad \exists y \in D \quad \text{such that} \quad x + y = 0.$$

Returning to *division*, the problem is to give a meaning, if possible, to expressions such as

$$\tfrac{1}{2}, \tfrac{3}{7}, \tfrac{2}{5}, \ \dots$$

We begin with $\tfrac{3}{7}$. To have any meaning at all, it must be expressible as one of the numbers of D: say $\exists x \in D$ such that

$$\tfrac{3}{7} = x.$$

By natural extension of the ideas of ordinary arithmetic, we expect this to be equivalent to the relation

$$3 = 7x,$$

and substitution of the nine (non-zero) possible values of x gives $x = 9$ as the unique solution. Thus *we may use the interpretation*

$$\tfrac{3}{7} = 9.$$

Consider next the symbol $\tfrac{1}{2}$. We require x such that

$$1 = 2x.$$

But this is *impossible*, since the left-hand side is odd and the right-hand side even. Hence

$$\nexists x \in D \quad \text{such that} \quad \tfrac{1}{2} = x;$$

that is, *the symbol $\tfrac{1}{2}$ has no meaning*.

To illustrate the dangers of unwary walking in digital arithmetic, consider next the symbol $\tfrac{2}{4}$ (which, to our 'ordinary arithmetic' eyes looks extremely like $\tfrac{1}{2}$). Here we require x such that

$$2 = 4x,$$

and inspection reveals *two* possible interpretations, namely $x = 3$ and $x = 8$. Thus *whereas we do not give any meaning within D to the symbol $\tfrac{1}{2}$, we have two choices, namely 3 and 8, for the symbol $\tfrac{2}{4}$*.

We have here a radical difference between ordinary arithmetic and digital arithmetic, and the cause is easily located. Take a slightly more significant example:

If

$$\tfrac{6}{8} = x,$$

then $x = 2$ or $x = 7$, whereas if

$$\tfrac{3}{4} = x,$$

then no value of x can be found.

Now the reason why, in ordinary arithmetic, we equate $\tfrac{6}{8}$ with $\tfrac{3}{4}$ is the *unique factorisation theorem* which states that *any given number N can be resolved uniquely, apart from order, into prime factors*.

For example, we have, uniquely,

$$6 = 2 \times 3,$$

$$8 = 2 \times 2 \times 2,$$

so that $$\frac{6}{8} = \frac{2 \times 3}{2 \times 2 \times 2} = \frac{3}{2 \times 2} = \frac{3}{4}.$$

For digital arithmetic, however, the argument breaks down—thereby revealing that the 'unique factorisation' of ordinary arithmetic does require proof since the properties of ordinary numbers must be in some way bound up with it. In fact, this whole conception of prime factors needs re-examination for digital arithmetic, as we shall see almost immediately.

EXAMPLES

1. Verify that, whenever the symbol

$$\frac{a}{2a} \quad (a \in D, \; a \neq 0),$$

has a meaning (in the sense just described) then in each case its values can be either 3 or 8.

2. Prove that the symbols

$$\tfrac{1}{3}, \; \tfrac{2}{3}$$

have unique interpretations.

3. Verify the relation

$$\tfrac{7}{3} = 2 + \tfrac{1}{3}$$

by direct computation of each side.

Examine the relation

$$\tfrac{8}{6} = 1 + \tfrac{2}{6}.$$

It may be helpful at this point to construct the *multiplication table* for digital arithmetic (omitting multiplication by zero):

	1	2	3	4	5	6	7	8	9
1	1	2	3	4	5	6	7	8	9
2	2	4	6	8	0	2	4	6	8
3	3	6	9	2	5	8	1	4	7
4	4	8	2	6	0	4	8	2	6
5	5	0	5	0	5	0	5	0	5
6	6	2	8	4	0	6	2	8	4
7	7	4	1	8	5	2	9	6	3
8	8	6	4	2	0	8	6	4	2
9	9	8	7	6	5	4	3	2	1

This table has several interesting features, to which we shall draw attention in the next chapters. For the moment, observe that, if by *prime* we mean a number that cannot be expressed as a product of two other numbers (excluding itself and unity), then *there are no primes in digital arithmetic*. For example,

$$3 = 7 \times 9,$$
$$5 = 3 \times 5 = 5 \times 5 = 7 \times 5 = 9 \times 5,$$
$$7 = 3 \times 9.$$

Any argument based on prime factors thus does not even begin, and features of ordinary arithmetic based upon them may be expected to undergo many variations.

CHAPTER II

POLYNOMIALS IN
DIGITAL ARITHMETIC

❡ 1. We have already seen (p. 18) that quadratic polynomials in digital arithmetic have surprising properties
as regards the number of their zeros. It seems therefore
worth while to look in further detail at some more general
polynomials.

Consider, as an example, the polynomial

$$4x^5 + 5x^4 + 5x^2 + 6x.$$

It may be expected to take ten values, one for each
integer in the set

$$D \equiv \{0, 1, 2, 3, 4, 5, 6, 7, 8, 9\}.$$

What is more unexpected is that *its value is always zero,
whatever the value of $x \in D$.* In ordinary arithmetic, of
course, for a polynomial of that type to be permanently
zero would be unthinkable.

EXAMPLE

Verify that $4x^5 + 5x^4 + 5x^2 + 6x = 0$
for all $x \in D$.

❡ 2. A TABLE OF POWERS

In order to proceed, we form a table giving the first
few powers of x for all non-zero values of $x \in D$.

x	x^2	x^3	x^4	x^5
1	1	1	1	1
2	4	8	6	2
3	9	7	1	3
4	6	4	6	4
5	5	5	5	5
6	6	6	6	6
7	9	3	1	7
8	4	2	6	8
9	1	9	1	9

The table exhibits the remarkable result that, *for all values of $x \in D$,*
$$x^5 \equiv x.$$

There is therefore no need to consider polynomials of degree greater than 4, since, for instance,

$$x^6 = x(x^5) = x^2,$$
$$x^7 = x^2(x^5) = x^3,$$
$$x^{12} = x^2(x^5)^2 = x^4.$$

¶ 3. DEDUCTIONS FROM THE TABLE; ROOTS OF THE INTEGERS

(i) *The square root $\sqrt{a}$*

Suppose that $a \in D$ and that we wish to evaluate its square root $\sqrt{a}$: that is, to find a number y such that

$$a = y^2.$$

The table shows that *this cannot be done if a has any of the values* 2, 3, 7, 8. Hence

$$\sqrt{2}, \ \sqrt{3}, \ \sqrt{7}, \ \sqrt{8}$$

do not exist.

The other numbers (except 5) have two square roots:

$$\sqrt{1} = 1 \text{ or } 9, \qquad \sqrt{6} = 4 \text{ or } 6,$$
$$\sqrt{4} = 2 \text{ or } 8, \qquad \sqrt{9} = 3 \text{ or } 7.$$

In each of these four cases, the sum of the roots is zero.

Finally,
$$\sqrt{5} = 5$$

only, there being just the one value.

(ii) *The cube root $\sqrt[3]{a}$.*

If, similarly, $\sqrt[3]{a} = y$, then

$$a = y^3$$

and the table shows that *each element of D has precisely one cube root.*

The reader who has dealt a little with complex numbers, and met ω, ω^2, the 'cube roots of unity' in ordinary complex arithmetic, may like to know where they have gone here:

If x is any cube root of unity, then

$$x^3 = 1,$$

or
$$x^3 - 1 = 0,$$

or
$$(x-1)(x^2 + x + 1) = 0.$$

Thus $x = 1$ (the normal solution) or else

$$x^2 + x + 1 \equiv 0.$$

Now *this equation cannot be solved within the set D*; for example, we may write it in the form

$$x^2 + x = -1 = 9,$$

so that
$$x(x+1) = 9.$$

The left-hand side is necessarily even and the right-hand side is necessarily odd; so there is no solution.

EXAMPLE

Prove further that the equation

$$(x-1)(x^2+x+1) = 0$$

has no solution arising from divisors of zero.

(iii) *The fourth root $\sqrt[4]{a}$.*

Solutions are now very restricted. If $\sqrt[4]{a} = y$, then $a = y^4$, so that a can have only the values 1, 6, 5. Then

$$\sqrt[4]{1} = 1, 3, 7, 9$$

and

$$\sqrt[4]{6} = 2, 4, 6, 8.$$

Finally,

$$\sqrt[4]{5} = 5$$

only.

The two sets $\{1, 3, 7, 9\}$ and $\{2, 4, 6, 8\}$ will assume considerable significance later.

❙ 4. THE POLYNOMIAL OF ❙ 1

Let us return to the polynomial

$$4x^5 + 5x^4 + 5x^2 + 6x.$$

Since x^5 is the same as x, this is

$$5x^4 + 5x^2 + (4+6)x = 5x^4 + 5x^2$$
$$= 5x^2(x^2+1).$$

Now $5x^2$ is zero if x is even and $5(x^2+1)$ is zero if x is odd. Hence

$$4x^5 + 5x^4 + 5x^2 + 6x \equiv 0$$

for all $x \in D$.

❙ 5. THE FACTORIAL FUNCTION

In digital arithmetic the factorial function becomes so simple as to be useless. Thus

$$1! = 1, \quad 2! = 2, \quad 3! = 6, \quad 4! = 4,$$

and

$$n! = 0 \quad (n > 4).$$

¶ 6. POLYNOMIAL GRAPHS

With only ten available points for plotting and only ten available values in any case, the graphs of polynomials are necessarily tenuous. In fact, each graph consists of precisely ten dots—save that, to complete the 'picture', we usually add the value $x = 10$, giving eleven in all. The three 'curves' $y = x^2$, $y = x^3$, $y = x^4$ are given below.

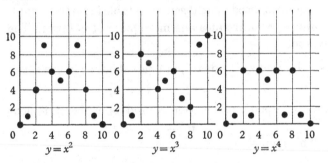

Fig. 1

EXAMPLE

Sketch the 'curves' for which

$$x^2 + y^2 = k$$

when $k = 0,\ 5,\ 7$.

CHAPTER III

THE IDEA OF A GROUP

⟦ 1. The multiplication table for digital arithmetic (p. 28) has several curious features which are worth considering both for themselves and also for the lead that they give into ideas of progressively greater abstraction.

Before proceeding to discuss the table in further detail, however, we introduce one more piece of notation. The symbol
$$\Rightarrow$$
is to be read as 'leads to', or equivalent, as in the statement
$$2x = 4 \quad \Rightarrow \quad 4x = 8;$$
the corresponding negation is used as in the statement
$$2x = 4 \quad \nRightarrow \quad 4x = 7.$$
The symbol
$$\Leftrightarrow$$
with the arrow-heads pointing both ways is to be read as 'leads to and arises from', as in the statement
$$3x = 7 \quad \Leftrightarrow \quad x = 9.$$
The negation is illustrated by the statement
$$3x = 7 \quad \nLeftrightarrow \quad x = 8.$$

Examples

Insert the appropriate symbol $\Rightarrow$, $\nRightarrow$, $\Leftrightarrow$ in place of $\sim$ in each of the following statements for numbers in digital arithmetic:

1. $5x(x+5) = 0 \qquad \sim \quad x = 7.$
2. $7x = 9 \qquad \sim \quad x = 7.$
3. $x^2 = x \qquad \sim \quad x = 0 \quad \text{or} \quad x = 1.$
4. $2x + 1 = 5x + 6 \quad \sim \quad x = 3.$

¶ 2. Properties of the Multiplication Table

Let us now look at some of the deductions that can be made by inspection of the table.

(i) If a is any one of the *odd* numbers 1, 3, 7, 9, then the row (or column) headed by it contains all the digits $1, 2, ..., 9$, each occurring once precisely. Thus the relation

$$ax = b \quad (b \in D \text{ also})$$

is satisfied by exactly one value of x, whatever value is selected for b.

For example,
$$3x = 8 \quad \Leftrightarrow \quad x = 6,$$
$$7x = 4 \quad \Leftrightarrow \quad x = 2,$$
$$9x = 6 \quad \Leftrightarrow \quad x = 4.$$

In terms of the earlier remarks about rational fractions, the symbol
$$\frac{b}{a}$$

has a unique interpretation within D when a is one of the numbers 1, 3, 7, 9.

(ii) If a is any one of the *even* numbers, then the row (or column) headed by it contains zero (corresponding to $5a$) and also the four numbers 2, 4, 6, 8 *each taken twice*. Thus the relation

$$ax = b \quad (b \in D)$$

has no solution when b is odd, and two solutions when b is even.

For example (and note the arrows carefully)
$$x = 6 \quad \Rightarrow \quad 4x = 4,$$
but
$$4x = 4 \quad \not\Rightarrow \quad x = 6;$$

$$x = 8 \quad \Rightarrow \quad 2x = 6,$$
but
$$2x = 6 \quad \not\Rightarrow \quad x = 8;$$
$$x = 4 \quad \Rightarrow \quad 8x = 2,$$
but
$$8x = 2 \quad \not\Rightarrow \quad x = 4.$$

In each case, the second line of the argument requires the symbol $\not\Rightarrow$, since another value of x is possible in addition to the one stated.

(iii) When $a = 5$, the relation

$$ax = b \quad (b \in D)$$

has no solution except when $b = 5$ or 0. (The value $b = 0$ has been excluded by implication from (i) and (ii).)

When $b = 5$, the equation

$$5x = 5$$

is satisfied by $x = 1, 3, 5, 7, 9$; when $b = 0$, the equation

$$5x = 0$$

is satisfied by $x = 0, \ 2, 4, 6, 8$.

¶ 3. THE EVEN INTEGERS 2, 4, 6, 8 UNDER DIGITAL MULTIPLICATION

Consider the *subset* of D formed by the elements 2, 4, 6, 8 taken from it. These elements form a set

$$E \equiv \{2, 4, 6, 8\}$$

which we may subject to the process of digital multiplication. The multiplication table (the order of the numbers being selected for later convenience) is:

	6	8	2	4
6	6	8	2	4
8	8	4	6	2
2	2	6	4	8
4	4	2	8	6

It is observed, and is very important for later work, that *each row and each column of the table contains each element of E once and once only.* The four numbers form a closed entity under digital multiplication.

An unexpected feature is that *the set E possesses a unity, namely the number* 6; for

$$6a = a$$

for all $a \in E$. This is an excellent example to warn us not to be led astray by notation or preconceived ideas (compare p. 24).

¶ 4. THE ODD INTEGERS 1, 3, 7, 9 UNDER DIGITAL MULTIPLICATION

Working similarly with the set

$$F \equiv \{1, 3, 7, 9\}$$

we have the table

	1	3	7	9
1	1	3	7	9
3	3	9	1	7
7	7	1	9	3
9	9	7	3	1

This set also has a unity, this time the number 1.

The point now to be made is that *these two tables are structurally the same*, in that if we form a more general table

	e	a	b	c
e	e	a	b	c
a	a	c	e	b
b	b	e	c	a
c	c	b	a	e

then the first is the special case

$$e = 6, \quad a = 8, \quad b = 2, \quad c = 4$$

and the second is the special case

$$e = 1, \quad a = 3, \quad b = 7, \quad c = 9.$$

¶ 5. THE $\{e, a, b, c\}$ TABLE AGAIN

It is impressive enough that the same multiplication table should serve two such apparently different uses, but many other particularisations are possible. We revert for a few moments to *ordinary arithmetic*. Take e to be the ordinary unity 1, and suppose that the four numbers e, a, b, c are all different. From the table (with $e = 1$) we have, for a use to be possible,

$$a^2 = c, \quad ab = 1, \quad ac = b, \quad b^2 = c, \quad bc = a, \quad c^2 = 1.$$

Since $c^2 = 1$ and $c \neq 1$ (being different from e), it follows that

$$c = -1.$$

Hence $$a^2 = -1, \quad b^2 = -1.$$

In elementary arithmetic, there are no possible values for a and b; but if we allow the introduction of *complex numbers*, then we can take

$$a = i \quad (= \sqrt{(-1)})$$

and $b \ (\neq a)$ is then $-i$.

[The reader who has not yet met complex numbers will find it sufficient for the present to accept i as a number of ordinary algebra to be manipulated in accordance with the ordinary rules save only that i^2 may always be replaced by -1. For example,

$$(p + iq)(p - iq) = p^2 - (iq)^2 = p^2 - (-q^2)$$
$$= p^2 + q^2.]$$

If, then, we set

$$e = 1, \quad a = i, \quad b = -i, \quad c = -1,$$

we may hope that the multiplication table

	1	i	$-i$	-1
1	1	i	$-i$	-1
i	i	-1	1	$-i$
$-i$	$-i$	1	-1	i
-1	-1	$-i$	i	1

will be valid; and it is easy to check that this is so.

EXAMPLES

1. Prove that the same form of table as that given in
¶ 4 is obtained (when $e = 1$, $a = 2$, $b = 3$, $c = 4$) if a
'product' xy means the remainder after dividing the
ordinary arithmetical product of x and y by 5.

2. A set consists of the numbers 2, 4, 6, 8. The rule of
'multiplication' is that a 'product' xy means the units
digit of the ordinary arithmetical product $\frac{1}{2}xy$. Obtain
the multiplication table.

¶ 6. Since we have touched on complex numbers, it may
be worth while to point out their analogues in digital
arithmetic. We have seen earlier the interpretation

$$-1 = 9,$$

and so $\sqrt{(-1)}$ is any number which multiplied (digitally)
by itself gives 9. There are two such numbers, 3 and 7;
so, noting that their sum is zero, we obtain the interpretations
$$i = 3, \quad -i = 7.$$

This is, of course, what lies behind the essential identity
of the multiplication tables for the sets $\{1, 3, 7, 9\}$ and
$\{1, i, -i, -1\}$.

EXAMPLE

Prove that, under the rules of digital arithmetic, the set $\{2, 4, 6, 8\}$ has 6 as unity, 4 as '-1' and 2, 8 as '$\sqrt{(-1)}$', '$-\sqrt{(-1)}$' respectively.

¶ 7. THE IDEA OF A GROUP

Consider again the general table

	e	a	b	c
e	e	a	b	c
a	a	c	e	b
b	b	e	c	a
c	c	b	a	e

This involves four elements, among them the 'unity' element e, and sixteen products. It is easy to check that the products satisfy the *associative law* for multiplication: for example,

$$(ab)c = (e)c = c,$$

$$a(bc) = a(a) = a^2 = c;$$

and

$$(ba)b = (e)b = b,$$

$$b(ab) = b(e) = b;$$

and so on. So we have *four elements whose products satisfy the associative law*.

Further, *each row and each column contains each element once and once only*.

A *set* of elements subject to a *law of combination* ('multiplication' in an extended sense of the word) satisfying the above two conditions is said to form a *group*.

We have introduced the idea of a group from this somewhat pictorial point of view, since that seems to show quickly what is involved. In more theoretical treatments other properties are used for the actual definition. These will now be obtained, establishing them from the multiplication table.

First, however, we must remove a possible source of misconception. This particular group is a *commutative* group (also called an *Abelian* group) in that all pairs of elements commute for multiplication; for example we obtain the same answer for a product *ab* as for its commuted form *ba*. But this need not always be so. To give 'room for manoeuvre' we select as an example a more elaborate group,* of six elements, forming thirty-six 'products':

	e	a	b	u	v	w
e	e	a	b	u	v	w
a	a	b	e	w	u	v
b	b	e	a	v	w	u
u	u	v	w	e	a	b
v	v	w	u	b	e	a
w	w	u	v	a	b	e

It is necessary to explain carefully what is meant here by the word *product*. Take, for instance, the symbol *wa*. We mean by *wa* that element which is in the *row* of *w* and the *column* of *a*; thus

$$wa = u.$$

On the other hand, we mean by *aw* that element which is in the *row* of *a* and the *column* of *w*; thus

$$aw = v.$$

The two products are thus unequal; in other words,

$$wa \neq aw,$$

* This group will be derived from an argument in elementary geometry later; see p. 85. There is a slight change of notation.

and so we must be very careful about the order in which we write the elements of a product.

For further illustration, the table shows that

$$uv = a, \quad vu = b;$$

$$ua = v, \quad au = w.$$

The first of the requirements (p. 40) for the six elements to form a group is that their 'products' satisfy the *associative law* $x(yz) = (xy)z$ for all x, y, z in the given set:

EXAMPLES

Verify that the products from these elements do satisfy the *associative law*, confining your attention if you wish to the following examples:

(i) $(au)a = a(ua)$,

(ii) $(ab)u = a(bu)$,

(iii) $(uv)w = u(vw)$,

(iv) $(wb)v = w(bv)$.

Assuming, then, that the associative law has been established, as in the above examples, note that *the six elements subject to this multiplication table do form a group*, in the sense of the definition of p. 40. The aim now is to derive through this group the more abstract form of general definition to which we just referred. The argument will be based on a particular case, but the reasoning can be applied to any group with a given multiplication table.

¶ 8. FUNDAMENTAL GROUP PROPERTIES

For definiteness, the following argument, which is perfectly general in nature, refers throughout to the group table for the six elements e, a, b, u, v, w described in ¶ 7.

Note first that, *if three elements x, y, p are so related that*

$$xp = yp,$$

then x and y are the same element. For the column through p contains the element which is xp or yp once only, so that its row is determined. Similarly,

$$qx = qy \quad \Rightarrow \quad x = y.$$

We come now to the basic group properties:

1. *The group necessarily has a unity element.* By a unity element, it will be recalled, is meant an element denoted (without prejudice) by e such that, for all x in the group, $$ex = xe = x.$$

Take x as a typical element in the group. The column through x contains all the elements and, in particular, x itself. Hence $\exists f$ so that $fx = x$.

Similar argument from another typical element y leads to an element g such that

$$gy = y.$$

We cannot, without further argument, assume that f and g will be the same. But

$$fx = x$$
$$\Rightarrow \quad f(fx) = fx$$
$$\Rightarrow \quad (ff)x = fx$$
$$\Rightarrow \quad ff = f;$$

and
$$gy = y$$
$$\Rightarrow \quad f(gy) = fy$$
$$\Rightarrow \quad (fg)y = fy$$
$$\Rightarrow \quad fg = f.$$

Hence
$$ff = fg,$$
so that
$$f = g.$$

Hence $\exists f$, *the same for all elements*, such that

$$fx = x.$$

Similarly, $\exists h$, the same for all elements z, such that

$$zh = z.$$

We have now to prove that f and h are the same:

Take for x the element h, and for z the element f; thus

$$fh = h, \quad fh = f.$$

Hence
$$h = f.$$

We have therefore established the existence of an element, which we now call e, such that

$$ex = xe = x$$

for all x of the group.

(The particular group exhibits e at once from the table as a unity element, but merely to look at one group is hardly general argument.)

2. *For each element x of the group an element y can be found such that*
$$xy = yx = e.$$

The elements x, y are said to be *inverse* in the group. It is natural to write
$$y = x^{-1}$$

and easy to prove that
$$(x^{-1})^{-1} = x.$$

To prove the main result, the table establishes, for any given x, the existence of y and z such that
$$xy = e, \quad zx = e.$$
The problem is to prove that y and z are the same.

Now
$$xy = e$$
$$\Rightarrow z(xy) = ze = z$$
$$\Rightarrow (zx)y = z$$
$$\Rightarrow ey = z$$
$$\Rightarrow y = z.$$

To recapitulate: a *group* may be defined abstractly as a *set S* of elements subject to a *rule of combination*, which we denote by the symbol $\times$, such that
$$x \in S, \quad y \in S \quad \Rightarrow \quad x \times y \in S,$$
$$(x \times y) \times z = x \times (y \times z),$$
$$\exists e \quad \text{such that} \quad x \times e = e \times x = x,$$
$$\exists x^{-1} \quad \text{such that} \quad (x^{-1}) \times x = x \times (x^{-1}) = e,$$
where x, y, z are arbitrary elements of S.

This is the definition usually used as a starting-point for a theoretical discussion.

EXAMPLES

1. In the first table of ¶ 7, identify
$$a^{-1}, b^{-1}, c^{-1}.$$

2. In the second table of ¶ 7, identify
$$a^{-1}, b^{-1}, u^{-1}, v^{-1}, w^{-1}.$$

3. Prove that, if x, y are elements of any group S,

$$(xy)^{-1} = y^{-1}x^{-1}$$

for multiplication in the order stated.

NOTE. A final comment may be useful. All the groups to be considered here have a finite number of elements—four and six have been met so far. But it is perfectly possible to construct groups having an infinite number of elements: for example, the set $$\{\ldots, -3, -2, -1, 0, 1, 2, 3, \ldots\}$$ where the 'product' of two numbers a and b is the ordinary algebraic sum, so that

$$-3+2 = -1, \quad 6+(-4) = 2,$$

and so on; the 'unity' is the number 0 and the 'inverse' of an element a is the number $-a$.

⁋ 9. ANOTHER GROUP OF SIX ELEMENTS

Take the set of six numbers

$$1, 2, 3, 4, 5, 6$$

and subject them to the rule that the product ab is to mean the remainder after dividing the ordinary product $a \times b$ by 7. The table (where the order selected for the numbers is adopted for the sake of points to be illustrated later) is

	1	2	4	6	3	5
1	1	2	4	6	3	5
2	2	4	1	5	6	3
4	4	1	2	3	5	6
6	6	5	3	1	4	2
3	3	6	5	4	2	1
5	5	3	6	2	1	4

EXAMPLES

1. Prove that these numbers, subject to this rule, form a group.

2. Find the inverses of each of the six elements.

3. Prove that the three numbers 1, 2, 4, subject to the same rule, also form a group; and that the two numbers 1, 6, subject to the same rule, also form a group.

4. Prove that the above group and the other group of six elements given on p. 41 are quite distinct, in spite of obvious similarities.

5. The elements of a set are the six numbers $1, \omega, \omega^2$, $-1, -\omega^2, -\omega$, where ω (a 'cube root of unity') is subject to the normal rules of arithmetic save that $\omega^3 = 1$ although $\omega \neq 1$. Prove that these six numbers form, under multiplication, a group whose table has the same structure as that just given.

CHAPTER IV

A GROUP OF 'PRODUCT' OPERATIONS IN GEOMETRY

¶ 1. THE FIGURE

Let A, B, C, D be four points in general position in a plane; denote by P, Q, R the intersections

$$P \equiv (BC, AD), \quad Q \equiv (CA, BD), \quad R \equiv (AB, CD).$$

As a complete abstraction, which the reader is advised not to attempt to visualise, we introduce an *identification point E*, which is there simply to have the properties

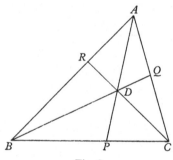

Fig. 2

with which we shall shortly endow it. (The beginner should not allow himself to be put off by this abstraction; all things will work together for good.)

¶ 2. THE OPERATIONS

We introduce eight 'operations', to be denoted by the symbols $e, p, q, r, d, a, b, c.$

(i) The operation e is quickly disposed of; it is the identity operation leaving all the points of the configuration unchanged.

(ii) The operations a, b, c, d are also fairly simple. The operation a, for example, interchanges the points on the lines collinear with A: thus it replaces B, C, D, P, Q, R by R, Q, P, D, C, B respectively. This process may conveniently be denoted by the symbolism

$$a(BCDPQR) = (RQPDCB).$$

Similarly,

$$b(CADPQR) = (PRQCDA),$$

$$c(ABDPQR) = (QPRBAD),$$

$$d(ABCPQR) = (PQRABC).$$

(iii) The operations p, q, r are a little harder to explain. Consider first the operation p. By analogy with the previous cases listed in (ii), it is to interchange B, C and also A, D. In addition, it is to interchange Q, R *as it would have done if P, Q, R had been a straight line.* Thus, by definition,

$$p(ABCDQR) = (DCBARQ),$$

$$q(ABCDRP) = (CDABPR),$$

$$r(ABCDPQ) = (BADCQP).$$

(iv) In addition, any operator acting on the point denoted by its own letter will be regarded as interchanging that point with the identification point E; for example, as a matter of definition,

$$a(AE) = (EA),$$

$$p(PE) = (EP),$$

and so on.

⟪ 3. THE 'PRODUCT' OF TWO OPERATIONS

Suppose that the eight points $EPQRDABC$ are subjected to the operation a; thus

$$a(EPQRDABC) = (ADCBPERQ).$$

Subject the new sequence of points

$$ADCBPERQ$$

to the operation p; thus

$$p(ADCBPERQ) = (DABCEPQR).$$

In obvious sense of the notation,

$$p\{a(EPQRDABC)\} = (DABCEPQR),$$

and it is natural to write the left-hand side in the form

$$pa(EPQRDABC).$$

The effect is to act on $(EPQRDABC)$ by an operation conveniently denoted by the symbol pa. This operation is called the *product* of a by p.

Note carefully that the product of p by a is

$$ap(EPQRDABC)$$
$$= a\{p(EPQRDABC)\}$$
$$= a(PERQADCB)$$
$$= (DABCEPQR).$$

The operations pa and ap involve a and p in reverse orders. In this particular case the results have been the same, but *it need not be so*: in this kind of 'multiplication' it must not be assumed that two products uv and vu are necessarily the same.

As a second example, form the two products qa and aq. We have

$$qa(EPQRDABC) = q(ADCBPERQ)$$

$$= (CBADRQPE)$$

and $$aq(EPQRDABC) = a(QREPBCDA)$$

$$= (CBADRQPE).$$

⁋ 4. THE OPERATIONS AS A CLOSED SYSTEM

Let us return to the formula

$$pa(EPQRDABC) = (DABCEPQR)$$

and consider the right-hand side. It is, in fact, identifiable quickly as $$d(EPQRDABC),$$

so that *the effects of the two operations pa and d are the same*. In symbolic language,

$$pa = ap = d.$$

Similarly, $$qa = aq = c.$$

EXAMPLES

Prove the formulae:

(i) $a^2 = e$, $p^2 = e$, $d^2 = e$.

(ii) $bc = p$, $ca = q$, $ab = r$.

(iii) $pd = a$, $qd = b$, $rd = c$.

¶ 5. A 'MULTIPLICATION TABLE' OF OPERATIONS

It will be simpler to write the table down and to explain it after:

	e	p	q	r	d	a	b	c
e	e	p	q	r	d	a	b	c
p	p	e	r	q	a	d	c	b
q	q	r	e	p	b	c	d	a
r	r	q	p	e	c	b	a	d
d	d	a	b	c	e	p	q	r
a	a	d	c	b	p	e	r	q
b	b	c	d	a	q	r	e	p
c	c	b	a	d	r	q	p	e

The table is to be interpreted as follows:

To identify a product such as

$$ap,$$

find the element in the same *row* as a and in the same *column* as p: that is, associate row with a and column with p. (Since all these products are, as it happens, commutative, the insistence on row-and-column distinction is superfluous; but there are many cases where the distinction is vital; see p. 41.)

EXAMPLE

Verify that the product of two operations, such as aq, is found by obtaining the third point (C) on the line AQ, giving the operation c. The points P, Q, R are treated as collinear for this purpose.

¶ 6. THE ASSOCIATIVE LAW

The associative law, for example

$$a(qd) = (aq)d$$

follows immediately from the definition of the operations, each side being in fact the result of operating first by d, then by q, and then by a. The result can also be verified, somewhat laboriously, from the table:

$$a(qd) = ab = r,$$
$$(aq)d = cd = r.$$

EXAMPLES

Verify the relations:

$$a(bc) = (ab)c,$$
$$b(rb) = (br)b,$$
$$q(pd) = (qp)d.$$

¶ 7. THE GROUP PROPERTY

It follows at once from the discussion on pp. 40–45 that *the set of elements subject to the multiplication table given in ¶ 5 forms a group.*

¶ 8. A SUBGROUP

The top right-hand corner of the table given in ¶ 5 is

	e	p	q	r
e	e	p	q	r
p	p	e	r	q
q	q	r	e	p
r	r	q	p	e

and brief inspection shows that *this is a group in its own right.* Such a section of a given group is called a *subgroup.*

EXAMPLES

1. Prove that the following are also subgroups:

(i) *adpe*,   (ii) *bpce*.

2. Prove that the following are not subgroups:

(i) $dabc$,   (ii) $dpqr$.

¶ 9. A COMPARISON OF GROUPS

We have now obtained two groups of four elements each, namely (p. 37)

	e	a	b	c
e	e	a	b	c
a	a	c	e	b
b	b	e	c	a
c	c	b	a	e

and (p. 53)

	e	p	q	r
e	e	p	q	r
p	p	e	r	q
q	q	r	e	p
r	r	q	p	e

It is natural to ask whether they are in fact the same group, apart from notation. The answer is immediately *no*, since all elements of the second group have their 'squares' equal to the unity, whereas there are two elements of the first whose 'squares' are not unity. Hence the two groups are quite distinct.

¶ 10. ANOTHER WAY TO THE SAME GROUP OF EIGHT ELEMENTS

A great deal of the interest of the group in chapter III arose from the fact that several different approaches led to the same end. In a similar way, we now give an entirely different approach to the group of eight elements considered in this chapter.

By an *ordered triplet* we shall mean a set of three numbers written in the form

$$(a, b, c),$$

where the order is important so that, in general,

$$(a, b, c) \neq (b, a, c) \neq (b, c, a).$$

The numbers themselves will be calculated in *binary arithmetic*, which is very similar to the digital arithmetic of chapter I, save that we require only the operation of addition. By the *sum* of two numbers a and b we mean the remainder on dividing the normal sum $a+b$ by 2. Thus
$$7+5 = 0, \quad 9+4 = 1.$$

In fact, however, only the two elements 0, 1 are required, and they are subject to the four rules

$$0+0 = 0, \quad 0+1 = 1, \quad 1+0 = 1, \quad 1+1 = 0.$$

The available triplets, to which we now give names, are

$$e \equiv (0,0,0), \quad p \equiv (0,1,1), \quad q \equiv (1,0,1), \quad r \equiv (1,1,0),$$
$$d \equiv (1,1,1), \quad u \equiv (1,0,0), \quad v \equiv (0,1,0), \quad w \equiv (0,0,1).$$

To combine them, we use the word 'product', where, as a matter of definition, the *product* of the two elements

$$x \equiv (a, b, c), \quad y \equiv (l, m, n)$$

is to be given by the formula

$$xy \equiv (a+l, b+m, c+n),$$

the addition on the right being in accordance with binary arithmetic as just described.

For example,

$$pq = (0+1, 1+0, 1+1) = (1, 1, 0),$$
$$dr = (1+1, 1+1, 1+0) = (0, 0, 1),$$
$$uw = (1+0, 0+0, 0+1) = (1, 0, 1).$$

EXAMPLES

Find the products:

(i) pr,　(ii) qu,　(iii) pd,　(iv) wp.

It is now a simple matter to construct the table:

	e	p	q	r	d	u	v	w
e	e	p	q	r	d	u	v	w
p	p	e	r	q	u	d	w	v
q	q	r	e	p	v	w	d	u
r	r	q	p	e	w	v	u	d
d	d	u	v	w	e	p	q	r
u	u	d	w	v	p	e	r	q
v	v	w	d	u	q	r	e	p
w	w	v	u	d	r	q	p	e

which is identical with that given on p. 52.

Hence *the two systems described in this chapter have identical group structures.*

¶ 11. YET ANOTHER WAY TO THE SAME GROUP

(Note the symmetry of notation used to explain Fig. 3.)

Let $O1$, $O2$, $O3$ be three given mutually perpendicular lines in space. Denote by α, β, γ (Greek letters alpha, beta, gamma) the three planes, also mutually perpendicular, $O23$, $O31$, $O12$.

Given any point E, we define eight operations:

(i) eE, meaning, 'leave E alone';

(ii) aE, meaning, 'take the mirror image of E in the plane α'—that is, take the point A such that EA is perpendicular to the plane α and bisected by it;

(iii) bE, meaning, 'take the mirror image of E in the plane β';

(iv) cE, meaning, 'take the mirror image of E in the plane γ';

(v) dE, meaning, 'take the mirror image of E in the point O'—that is, take the point D such that ED is bisected by O;

(vi) uE, meaning, 'take the mirror image of E in the line $O1$'—that is, take the point U such that EU is perpendicular to the line $O1$ and bisected by it;

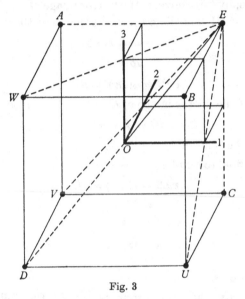

Fig. 3

(vii) vE, meaning, 'take the mirror image of E in the line $O2$';

(viii) wE, meaning, 'take the mirror image of E in the line $O3$'.

In the diagram (Fig. 3) these points are $E, A, B, C, D,$ U, V, W respectively.

We now require a meaning for the *product* of two or more of these operations.

Consider, for example, the meaning to be given to the symbol
$$ubE.$$

To make sense at all, this must be
$$u(bE),$$
or
$$uB;$$
and this, by (vi) above, is the mirror image *of the point B* in the line $O1$. The diagram shows this to be the point C. Hence
$$ubE = C = cE,$$
so that
$$ub = c$$

Again,
$$acE = a(cE) = aC,$$
which is the mirror image of C in the plane α; so that
$$aC = V = vE,$$
giving the relation
$$ac = v.$$

As a final example,
$$vuE = v(uE) = vU$$
$$= W$$
$$= wE,$$
so that
$$vu = w.$$

EXAMPLES

1. Verify that the products satisfy the following table:

	e	u	v	w	d	a	b	c
e	e	u	v	w	d	a	b	c
u	u	e	w	v	a	d	c	b
v	v	w	e	u	b	c	d	a
w	w	v	u	e	c	b	a	d
d	d	a	b	c	e	u	v	w
a	a	d	c	b	u	e	w	v
b	b	c	d	a	v	w	e	u
c	c	b	a	d	w	v	u	e

2.   Verify that these operations form a group identical in structure with those given on pp. 52 and 56.

3.   Verify that a set of elements e, a, b, c satisfying any 'multiplication table'

	e	a	b	c
e	a	e	c	b
a	e	c	b	a
b	c	b	a	e
c	b	a	e	c

cannot form a group, there being no unity element and the associative law not being obeyed.

CHAPTER V

GROUPS WITHIN GROUPS

We have now advanced some distance in abstraction. The next two chapters will give a gentle introduction to some standard results that follow fairly naturally, presented here from a more or less intuitive point of view.

¶ 1. THE GROUPS DISPLAYED

We gather together, with occasional changes of notation or of the order in which elements are written, the groups that we have already met (including some 'subgroups'):

A (p. 47)

	e	a
e	e	a
a	a	e

B (p. 47)

	e	a	b
e	e	a	b
a	a	b	e
b	b	e	a

C (p. 37). The meaning of the dotted lines in the following tables will appear later.

	e	a	b	c
e	e	a	b	c
a	a	e	c	b
b	b	c	a	e
c	c	b	e	a

D (p. 53)

	e	a	b	c
e	e	a	b	c
a	a	e	c	b
b	b	c	e	a
c	c	b	a	e

E (p. 41)

	e	a	b	u	v	w
e	e	a	b	u	v	w
a	a	b	e	w	u	v
b	b	e	a	v	w	u
u	u	v	w	e	a	b
v	v	w	u	b	e	a
w	w	u	v	a	b	e

F (p. 46)

	e	a	b	u	v	w
e	e	a	b	u	v	w
a	a	b	e	w	u	v
b	b	e	a	v	w	u
u	u	w	v	e	b	a
v	v	u	w	b	a	e
w	w	v	u	a	e	b

G (p. 52)

	e	a	b	c	d	u	v	w
e	e	a	b	c	d	u	v	w
a	a	e	c	b	u	d	w	v
b	b	c	e	a	v	w	d	u
c	c	b	a	e	w	v	u	d
d	d	u	v	w	e	a	b	c
u	u	d	w	v	a	e	c	b
v	v	w	d	u	b	c	e	a
w	w	v	u	d	c	b	a	e

¶ 2. GROUPS OF TWO OR THREE ELEMENTS

Here, as always throughout this chapter, we use the symbol e for the unity element. We have seen (p. 43) that every group must have a unity.

Consider first a group of *two* elements, e and a. The group table necessarily starts in the form

	e	a
e	e	a
a	a	.

The only uncertain product is indicated by a dot; it must, in fact, be e, for the elements in rows (or columns) must be distinct.

Hence *the only table for a group of two elements is*

	e	a
e	e	a
a	a	e

Corollary. The element a satisfies the relation

$$a^2 = e.$$

Consider next a group of *three* elements e, a, b. The table begins

	e	a	b
e	e	a	b
a	a	.	.
b	b	.	.

The second row can thus be, in the first place, either a, e, b or a, b, e. The former is, however, impossible, since under it the third column would have two elements b. The whole table is now determined, so that *the only table for a group of three elements is*

	e	a	b
e	e	a	b
a	a	b	e
b	b	e	a

Corollary. The elements a, b satisfy the relations

$$a^2 = b, \quad b^2 = a.$$

EXAMPLES

1. Prove that
$$a^3 = e, \quad b^3 = e,$$
$$a^{-1} = b, \quad b^{-1} = a,$$

where a^{-1}, b^{-1} are the inverses (p. 44) of a, b.

2. Establish for group C the relations

$$b^4 = e, \quad c^4 = e.$$

¶ 3. SUBGROUPS

In order to examine the structure of a given group, one obvious starting-point is to consider whether there are within the group any *subgroups*: sets of elements, that is, within the group which, under the given rule for group multiplication, have by themselves all the properties of a group.

Examples, some of which have been met already, are:

$$\{e, a\} \quad \text{in} \quad C;$$
$$\{e, a\} \quad \text{in} \quad D;$$
$$\{e, a, b\}, \{e, u\}, \{e, v\}, \{e, w\} \quad \text{in} \quad E;$$
$$\{e, a, b\}, \{e, u\} \quad \text{in} \quad F.$$

Note that, just as a given group must contain a unity element, so *the unity e must be a member of every subgroup*.

EXAMPLE

Verify that if x is any element of a given group, then its inverse x^{-1} in that group is a member of every subgroup containing x.

⟦ 4. COSETS

Once a subgroup has been selected within a given group, an obvious question to ask is what features there are for the products that arise when *elements of the subgroup* are multiplied by *the elements of the whole group*.

Consider two examples:

(i) In group E, select the subgroup $\{e, a, b\}$. Then, in obvious notation:

$$\{e, a, b\}e = \{e^2, ae, be\} = \{e, a, b\},$$
$$\{e, a, b\}a = \{ea, a^2, ba\} = \{a, b, e\},$$
$$\{e, a, b\}b = \{eb, ab, b^2\} = \{b, e, a\},$$
$$\{e, a, b\}u = \{eu, au, bu\} = \{u, w, v\},$$
$$\{e, a, b\}v = \{ev, av, bv\} = \{v, u, w\},$$
$$\{e, a, b\}w = \{ew, aw, bw\} = \{w, v, u\}.$$

(ii) In group G, select the subgroup $\{e, c\}$. Then (expressed more briefly):

$$\{e, c\}e = \{e, c\}, \quad \{e, c\}d = \{d, w\},$$
$$\{e, c\}a = \{a, b\}, \quad \{e, c\}u = \{u, v\},$$
$$\{e, c\}b = \{b, a\}, \quad \{e, c\}v = \{v, u\},$$
$$\{e, c\}c = \{c, e\}, \quad \{e, c\}w = \{w, d\}.$$

The resulting 'product' sets divide themselves exclusively into classes. In (i), the classes are

$$\{e, a, b\}, \quad \{u, v, w\};$$

in (ii), the classes are

$$\{e, c\}, \quad \{a, b\}, \quad \{d, w\}, \quad \{u, v\}.$$

These sets are called the *cosets arising from the given subgroups*.

EXAMPLES

1. Prove that, in C, the cosets arising from the subgroup $\{e, a\}$ are $\{e, a\}$, $\{b, c\}$.

2. Prove that, in F, the cosets arising from the subgroup $\{e, u\}$ are $\{e, u\}$, $\{a, w\}$, $\{b, v\}$.

3. Prove that, in G, the cosets arising from the subgroup $\{e, a, b, c\}$ are $\{e, a, b, c\}$, $\{d, u, v, w\}$.

¶ 5. SOME SPECIAL FEATURES OF THE GROUPS $C, \ldots, G$

The features which follow are common to many groups *but not to all*. Our immediate aim is to illustrate the way in which from a given group we may be able to obtain groups within it of even greater abstraction.

Let us begin with the group C. The table (p. 60) is exhibited as divided into four blocks, of which two contain the elements e, a and two the elements b, c. Write the sets $\{e, a\}$, $\{b, c\}$ in the form:

$$H \equiv \{e, a\}, \quad K \equiv \{b, c\},$$

where H, K are the cosets arising from the subgroup H.

Observe now that

(i) if two elements of H are multiplied, the result is an H;

(ii) if an element of H is multiplied by an element of K (in that order), the result is a K;

(iii) if an element of K is multiplied by an element of H (in that order), the result is a K;

(iv) if two elements of K are multiplied, the result is an H.

5

These results may be summarised by means of the table:

	H	K
H	H	K
K	K	H

where, for example, the product KH means precisely the operation described in (iii), yielding the answer K.

We thus have a *group of cosets*.

EXAMPLES

Repeat the argument to obtain the same 'multiplication table' from D, E, F and G, where, in the latter case, only the 'central' dividing lines are used.

From our given groups we have therefore succeeded in deriving other groups by an exceedingly abstract definition of the 'products'. The resulting group, having just the *two* elements H, K, is (¶ 2) necessarily of the form of group A, as is immediately clear.

A concrete representation of H, K in a particular geometrical setting is given in the next chapter.

EXAMPLE

Verify that the above products for H, K do in fact satisfy the *associative law* essential (p. 40) for a group.

We pass on to the group G, considering now the full subdivision into sixteen 'squares'. Write

$$H \equiv \{a, e\}, \quad K \equiv \{b, c\}, \quad L \equiv \{d, u\}, \quad M \equiv \{v, w\},$$

the cosets arising from the subgroup H. We define products by exact analogy with the simpler case just discussed. To give three examples in illustration:

(i) $K \times K \equiv$ the set (if any) whose elements consist of the product of one K by another K

$$\equiv \{b \times b,\, b \times c,\, c \times b,\, c \times c\}$$
$$\equiv \{e,\, a,\, a,\, e\},$$

and this is just the set H, the repetitions being irrelevant. Hence
$$K \times K = H.$$

(ii) $L \times M \equiv$ the set (if any) whose elements consist of the product of one L by one M,

$$\equiv \{dv,\, dw,\, uv,\, uw\}$$
$$\equiv \{b,\, c,\, c,\, b\}$$
$$\equiv K.$$

(iii) $M \times H \equiv$ the set (if any) whose elements consist of the product of one M by one H

$$\equiv \{va,\, ve,\, wa,\, we\}$$
$$\equiv \{w,\, v,\, v,\, w\}$$
$$\equiv M.$$

Proceeding in this way, we obtain the table

	H	K	L	M
H	H	K	L	M
K	K	H	M	L
L	L	M	H	K
M	M	L	K	H

which we recognise as identical in structure with table D.

Once again we have a *group* of cosets.

EXAMPLE

Verify that the sets H, K, L, M subject to the present rule of 'multiplication' satisfy all the conditions for a group.

❡ 6. A REMARK ON SUBGROUPS

It is readily observed that C, D, G all have as a 'subgroup' the group described as A. But it is important to realise, as remarked before (compare p. 59), that the sets

$$\{b, c\} \quad \text{in} \quad C,$$
$$\{b, c\} \quad \text{in} \quad D,$$
$$\{b, c\}, \{d, u\}, \{v, w\} \quad \text{in} \quad G$$

are *not subgroups*: they cannot have the features of a group since *they do not possess a unity element*.

In the same way, $\{e, a, b\}$

is a subgroup of E and F, but

$$\{u, v, w\}$$

is not; and $\{e, a, b, c\}$

is a subgroup of G, but

$$\{d, u, v, w\}$$

is not.

❡ 7. As we implied (p. 65) the work given in ❡ 5 needs care in application. We give, in fact, a group for which modification is necessary before the calculations can be applied:

A NEW GROUP OF EIGHT ELEMENTS

Suppose that (x, y) is a number-pair consisting of two numbers x, y *in an assigned order*. We apply first a group (as it will be seen to be) of four 'transformations', which give to the two elements the four possible choices of sign:

$$e(x, y) = (x, y), \qquad b(x, y) = (x, -y),$$
$$a(x, y) = (-x, y), \quad c(x, y) = (-x, -y).$$

Following earlier precedent (p. 50), a *product* such as ab is defined by the rule

$$ab(x, y) = a\{b(x, y)\}$$
$$= a(x, -y)$$
$$= (-x, -y),$$

the operation a changing the sign of the first element while leaving the second (in this case $-y$) unchanged. Thus

$$ab(x, y) = c(x, y),$$

so that

$$ab = c.$$

Similarly,

$$ca(x, y) = c\{a(x, y)\}$$
$$= c(-x, y)$$
$$= (x, -y)$$
$$= b(x, y),$$

so that

$$ca = b.$$

Proceeding in this way we have the table

	e	a	b	c
e	e	a	b	c
a	a	e	c	b
b	b	c	e	a
c	c	b	a	e

showing that the four operations, with this rule for forming 'products', constitute a group (compare p. 61).

We now add four further operations obtained from the preceding by *interchanging* the two elements; thus we write

$$h(x, y) = (y, x), \qquad v(x, y) = (y, -x),$$
$$u(x, y) = (-y, x), \quad w(x, y) = (-y, -x).$$

The products are perhaps a little more awkward to calculate. We give three typical examples:

(i)
$$au(x, y) = a\{u(x, y)\}$$
$$= a(-y, x)$$
$$= (y, x)$$
$$= h(x, y),$$

so that
$$au = h.$$

(ii)
$$ua(x, y) = u\{a(x, y)\}$$
$$= u(-x, y)$$
$$= (-y, -x),$$

since the operation u is the instruction, 'interchange the elements and change the sign of the new first'. Thus

$$ua(x, y) = w(x, y),$$

so that
$$ua = w.$$

Note that *in this case the two products au and ua are different*. The order of the operations is of vital importance.

(iii)
$$v^2(x, y) = v\{v(x, y)\}$$
$$= v(y, -x)$$
$$= (-x, -y),$$
$$= c(x, y),$$

so that
$$v^2 = c.$$

Proceeding similarly for other products, we have the table

	e	a	b	c	h	u	v	w
e	e	a	b	c	h	u	v	w
a	a	e	c	b	u	h	w	v
b	b	c	e	a	v	w	h	u
c	c	b	a	e	w	v	u	h
h	h	v	u	w	e	b	a	c
u	u	w	h	v	a	c	e	b
v	v	h	w	u	b	e	c	a
w	w	u	v	h	c	a	b	e

where, for example, a product

$$ua$$

is obtained as the element in the *row* through u and the *column* through a.

EXAMPLES

1. Verify the above table.

2. Prove that the eight elements subject to these rules do form a group.

We may now follow the procedure given in ¶ 5 (p. 65), taking the subgroup $\{e, a, b, c\}$ and setting

$$H \equiv \{e, a, b, c\}, \quad K \equiv \{h, u, v, w\}.$$

The table is

	H	K
H	H	K
K	K	H

as before.

But if we seek further division on the pattern of p. 66, with four sections of two rows each crossed by four sections of two columns each, then *the subsequent calcula-*

tion does not work. If, for instance, we take the cosets of $\{e, a\}$ and put

$$H \equiv \{e, a\}, \quad K \equiv \{b, c\}, \quad L \equiv \{h, u\}, \quad M \equiv \{v, w\},$$

we can indeed give meaning to the eight products

$$H \times H, \quad H \times K, \quad H \times L, \quad H \times M,$$
$$K \times H, \quad K \times K, \quad K \times L, \quad K \times M,$$

but *the other eight products do not exist*:

For example, to form

$$L \times K,$$

or

$$\{h, u\} \times \{v, w\}$$

we need the set of products

$$\{hv, hw, uv, uw\},$$

or

$$\{a, c, e, b\},$$

which is *not* one of H, K, L, M. The subgroup $\{e, a\}$ thus does not give rise to a 'derived' group of the four elements H, K, L, M.

On the other hand, it is possible to obtain a 'derived' group of four members by rearranging the order of the given elements with respect to the subgroup $\{e, c\}$:

	e	c	u	v	h	w	a	b
e	e	c	u	v	h	w	a	b
c	c	e	v	u	w	h	b	a
u	u	v	c	e	a	b	w	h
v	v	u	e	c	b	a	h	w
h	h	w	b	a	e	c	v	u
w	w	h	a	b	c	e	u	v
a	a	b	h	w	u	v	e	c
b	b	a	w	h	v	u	c	e

Starting from the subgroup $\{e, c\}$ and writing

$$H \equiv \{e, c\}, \quad K \equiv \{u, v\}, \quad L \equiv \{h, w\}, \quad M \equiv \{a, b\},$$

we can obtain the group

	H	K	L	M
H	H	K	L	M
K	K	H	M	L
L	L	M	H	K
M	M	L	K	H

But *this group is based on* $\{e, c\}$ *and not on* $\{e, a\}$.

EXAMPLE

Verify the table of products for H, K, L, M and check that the elements do form a group.

¶ 8. It is not our aim* to get involved with the technical abstractions of group theory; the reader, however, will naturally wonder why the procedure just outlined gives groups in some cases but not in others. A break-down of the group structure will help to make this clearer, beginning with an example where the procedure has been seen to be successful.

Consider the group G (p. 61) and select from it any *subgroup*. Here, we choose, of course, the one adopted in the earlier calculation, namely

$$H \equiv \{e, a\}.$$

The first step is to multiply these two elements by each

---

* And, indeed, the reader may omit the remainder of this section if preferred.

member of the group in turn. In detail, calculations of the cosets give:

$$He \equiv \{e, a\}e \equiv \{ee, ae\} \quad \equiv \{e, \ a\},$$
$$Ha \equiv \{e, a\}a \equiv \{ea, aa\} \quad \equiv \{a, \ e\},$$
$$Hb \equiv \{e, a\}b \equiv \{eb, ab\} \quad \equiv \{b, \ c\},$$
$$Hc \equiv \{e, a\}c \equiv \{ec, ac\} \quad \equiv \{c, \ b\},$$
$$Hd \equiv \{e, a\}d \equiv \{ed, ad\} \quad \equiv \{d, \ u\},$$
$$Hu \equiv \{e, a\}u \equiv \{eu, au\} \quad \equiv \{u, \ d\},$$
$$Hv \equiv \{e, a\}v \equiv \{ev, av\} \quad \equiv \{v, \ w\},$$
$$Hw \equiv \{e, a\}w \equiv \{ew, aw\} \equiv \{w, v\},$$

and so the cosets arising from $\{e, a\}$ are (compare p. 66)

$$H \equiv \{e, a\}, \quad K \equiv \{b, c\}, \quad L \equiv \{d, u\}, \quad M \equiv \{v, w\}.$$

We have at once the formulae

$$H \times H = H, \quad H \times K = K, \quad H \times L = L, \quad H \times M = M$$

already used; the analogous four

$$H \times H = H, \quad K \times H = K, \quad L \times H = L, \quad M \times H = M$$

follow similarly.

The four elements H, K, L, M have thus arisen in a natural manner from a systematic calculation, and the first row and the first column of the new group table are verified:

	H	K	L	M
H	H	K	L	M
K	K			
L	L			
M	M			

At this stage of the argument we pause for a while.

The next step is to make (with greater brevity of exposition) the similar calculation with the 'unsuccessful' group tabulated on p. 72, starting with the subgroup

$$H \equiv \{e, a\}.$$

Then
$$He \equiv \{e, a\}, \quad Hh \equiv \{h, u\},$$
$$Ha \equiv \{a, e\}, \quad Hu \equiv \{u, h\},$$
$$Hb \equiv \{b, c\}, \quad Hv \equiv \{v, w\},$$
$$Hc \equiv \{c, b\}, \quad Hw \equiv \{w, v\}.$$

We find again the cosets

$$H \equiv \{e, a\}, \quad K \equiv \{b, c\}, \quad L \equiv \{h, u\}, \quad M \equiv \{v, w\},$$

and the results seem similar so far—the cosets of H are identical to the eye.

In order to trace the trouble to its source, we now proceed (still in the 'unsuccessful' case) to examine the product
$$L \times K.$$

This consists of the products obtainable from the cosets

$$\{h, u\}, \quad \{b, c\}$$

and so consists, in the first instance, of the four elements

$$\{hb, hc, ub, uc\}.$$

For the 'multiplication' to have a meaning, these four elements must in fact consist of *two only*, namely the elements of one or other of the cosets H, K, L, M. Now we cannot have $hb = hc$, since h is not the unity, nor $hb = ub$, since b is not the unity; hence we require, for 'success'
$$hb = uc$$

and, similarly,
$$hc = ub.$$

Consider, say, the first of these relations. We are basing the whole division of the given group upon the subgroup $\{e, a\}$ and its cosets, so it seems natural to begin by expressing some or all of the elements h, b, u, c in terms of a. Now from the table (p. 71), we have the relations

$$h = au, \quad c = ab,$$

so that

$$hb = uc$$

$$\Rightarrow aub = uab$$

$$\Rightarrow au = ua.$$

In other words, *the expression of the elements of the 'product' $L \times K$ in terms of one or other of H, K, L, M requires that the element u commutes with a.*

EXAMPLES

1. Prove similarly the requirement

$$ah = ha.$$

2. Prove, by considering other products, that, for the complete table with H, K, L, M to exist, it is necessary that, for all x in the given group,

$$ax = xa.$$

3. By consulting the table on p. 71, deduce that, in the 'unsuccessful' case, the elements H, K, L, M certainly cannot form a group.

The upshot of the preceding work is that, for

$$H, K, L, M$$

to form a group, it is *necessary* that

$$ax = xa$$

for all x in the given group. There is another way of expressing this: if x^{-1} is the inverse (p. 44) of x, so that

$$x(x^{-1}) = (x^{-1})x = e,$$

then
$$ax = xa$$
$$\Rightarrow x^{-1}ax = x^{-1}(xa)$$
$$= (x^{-1}x)a$$
$$= ea$$
$$= a.$$

That is, *the element a of a 'successful' subgroup* (on which to base the subdivision of the given group) *satisfies the relation*
$$x^{-1}ax = a$$

for all x in the given group. Thus *the element a is unaltered if multiplied before by x^{-1} and behind by x* If it *is* altered, the calculation will not work.

¶ 9. Suppose now that, *conversely*, there is an element a with this property, that

$$x^{-1}ax = a$$

for all x. Then *we can prove that the sets calculated from* $\{e, a\}$ *as on p.* 74 *do form a group.* Take, for example, the two sets formed from elements x, y of the given group. The two sets are (compare p. 74)

$$\{e, a\}x \equiv \{ex, ax\} \equiv \{x, ax\}$$
and
$$\{e, a\}y \equiv \{ey, ay\} \equiv \{y, ay\}.$$

Call these X, Y respectively. Then, by definition,

$$X \times Y$$

is the set whose elements are

$$\{xy, xay, axy, axay\}.$$

Now
$$x^{-1}ax = a$$
$$\Rightarrow ax = xa$$
$$\Rightarrow axy = xay,$$

so that the elements axy and xay are the same. Further,

$$x^{-1}ax = a$$
$$\Rightarrow ax = xa$$
$$\Rightarrow a^2x = axa;$$

but (p. 73), the whole calculation is based on the assumption that $\{e, a\}$ is a subgroup and therefore a group of two elements only, and there is thus a relation

$$a^2 = e.$$

Hence
$$x = axa,$$
so that
$$xy = axay,$$

and the two elements xy and $axay$ are therefore the same. The product $X \times Y$ thus consists of the *two* elements

$$\{xy, axy\}$$
$$\equiv \{e, a\}xy,$$

derived from $\{e, a\}$ on multiplication by that element of the given group which is the product of x, y.

We examine, finally, all the sets of the form

$$\{e, a\}x$$

for all x in the given group. They are equal in pairs, since the sets $\{e, a\}x$ and $\{e, a\}ax$ are the same, namely

$$\{e, a\}x \equiv \{x, ax\}$$
and
$$\{e, a\}ax \equiv \{ax, a^2x\}$$
$$\equiv \{ax, x\} \quad (a^2 = e).$$

They can therefore be grouped after the pattern

$$X \equiv \{e, a\}x \equiv \{e, a\}ax,$$
$$Y \equiv \{e, a\}y \equiv \{e, a\}ay,$$
$$Z \equiv \{e, a\}z \equiv \{e, a\}az,$$

and so on. By what we have just done, these sets may be subjected to the 'rule of multiplication'

$$X \times Y \equiv \{e, a\}xy,$$

which is clearly defined.

Now the set $\{X, Y, Z, \ldots\}$ under this rule has a 'unity'

$$\{e, a\}e;$$

and each element such as

$$\{e, a\}x$$

has an inverse $\qquad \{e, a\}x^{-1}.$

Subject only to confirmation of the associative law (p. 40) the elements therefore form a group; and the rule

$$([\{e, a\}x] \times [\{e, a\}y]) \times [\{e, a\}z]$$
$$= [\{e, a\}x] \times ([\{e, a\}y] \times [\{e, a\}z])$$

is an automatic consequence of the rule for 'multiplication' together with the associative law

$$(xy)z = x(yz)$$

for the given group.

The elements $X, Y, Z, \ldots$ therefore belong to a group.

¶ 10. The argument of ¶¶ 8, 9 has been cast into that particular form so as to exhibit how the expression

$$x^{-1}ax$$

must arise in order for the theory to develop. The case

when the subgroup, instead of two elements such as $\{e, a\}$, consists of four elements such as $\{e, a, b, c\}$ is more difficult, though the general principles are similar. A textbook on group theory should be consulted.

EXAMPLES

1. Prove for the form of group table given on p. 72 that the element c of the subgroup $\{e, c\}$ satisfies the relation

$$x^{-1}cx = c$$

for all x in the given group.

2. Prove that, if the given group is commutative (p. 41), then the rule '$x^{-1}ax = a$' always holds.

¶ 11. The point that we have reached, then, is that, *with care*, it may be possible from a given group to select one of its subgroups and then to define a new group by means of a definition of 'multiplication' involving considerable abstraction.

The chapter which follows gives geometrical flesh to some of these very abstract ideas.

CHAPTER VI

GROUP STRUCTURE ON A CIRCLE

I. A GROUP OF SIX POINTS

¶ 1. Given a straight line u, divide the 'straight angle' defined by it at any one of its points into three equal parts by the lines v, w (Fig. 4). The three directions u, v, w are to be the basic directions for subsequent work.

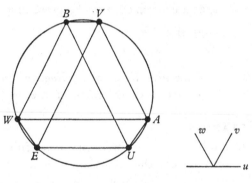

Fig. 4

In the plane of these lines, let a circle be given and take an arbitrary point E upon it. Five further points $U, A,$ V, B, W are constructed successively on the circle by means of the following operations:

(i) the line through E parallel to u cuts the circle again in U;

(ii) the line through U parallel to v cuts the circle again in A;

6

(iii) the line through A parallel to w cuts the circle again in V;

(iv) the line through V parallel to u cuts the circle again in B;

(v) the line through B parallel to v cuts the circle again in W.

(Since all the actual geometry is simple, we shall often carry the argument forward by means of *Examples* for the reader.)

EXAMPLE

Prove that, as a consequence of the preceding work:

(i) $WE \parallel w$;

(ii) $WA \parallel u$, $\quad EV \parallel v$, $\quad UB \parallel w$.

In other words, *the nine lines in the diagram fall into three sets of three each, in the directions of u, v, w.*

⟦ 2. NOTATION

Denote now by u, v, w the three *operations* of obtaining from any given point on the circle the second intersection (with the circle) of the line through that point parallel to the lines u, v, w respectively. Then the notation such as uB will be used to denote the result of operating on B by u; thus
$$uB = V.$$
In the same way, $\quad vV = E, \quad wB = U.$

The symbolism of 'products' follows naturally. Since
$$uE = U, \quad vU = A,$$
we write
$$v(uE) = A,$$
or
$$vuE = A.$$

That is, the operation denoted by vu consists of (i) acting on E by u, giving U; and then (ii) acting on the point so obtained by v, giving A.

Note that
$$uvE = u(vE)$$
$$= uV$$
$$= B,$$

so that *the products uv, vu are different*

These 'products' may be extended. For example,
$$uvwU = uv(wU)$$
$$= uvB$$
$$= u(vB)$$
$$= uW$$
$$= A;$$

and
$$vwuA = vw(uA)$$
$$= vwW$$
$$= v(wW)$$
$$= vE$$
$$= V.$$

It is important to note that *the 'products' obey the associative law* (p. 14).

EXAMPLE

Verify the equivalence of the operations
$$(uw)v = u(wv),$$
$$(vu)w = v(uw).$$

¶ 3. RELATIONS BETWEEN THE OPERATIONS

(i) We define first the *identity operator e* which has no effect upon the point on which it operates; thus, for example,
$$eE = E, \quad eV = V.$$

(ii) Suppose next that P is any one of the points E, A, B, U, V, W. Then, on operating by u, we have the relation
$$uP = Q,$$
say, where Q is that one of the six points for which $PQ \parallel u$. It follows that
$$uQ = P,$$
so that
$$u(uP) = P,$$
or
$$u^2P = eP$$
for all P. Hence *the operator u satisfies the relation*
$$u^2 = e.$$
Similarly
$$v^2 = e, \quad w^2 = e.$$

We therefore have *three fundamental identities*
$$u^2 = e, \quad v^2 = e, \quad w^2 = e.$$

(iii) We have seen that the 'product' operators uv, vu are different. What is perhaps more surprising is that the six operators obtained as products vw, wu, uv, wv, uw, vu can be grouped into two classes of three:
$$vwE = v(wE) = vW = B,$$
$$wuE = w(uE) = wU = B,$$
$$uvE = u(vE) = uV = B.$$
Thus
$$vw = wu = uv,$$
and, similarly,
$$wv = uw = vu.$$

It becomes natural to define new operators a, b by the relations

$$vw = wu = uv = b,$$

$$wv = uw = vu = a.$$

For example,

$$b^2 = (vw)(wu) = v(w^2)u = vu = a,$$

$$ba = (vw)(wv) = v(w^2)v = v^2 = e,$$

$$bu = (wu)u = w(u^2) = w.$$

¶ 4. THE 'PRODUCTS' TABLE

With the help of the three examples at the end of ¶ 3, and similar results, a table of 'products' may now be constructed. A 'product' such as

$$ab = e$$

is denoted by placing e in the *row* through a and the *column* through b. The table is:

	e	b	a	u	v	w
e	e	b	a	u	v	w
b	b	a	e	w	u	v
a	a	e	b	v	w	u
u	u	v	w	e	b	a
v	v	w	u	a	e	b
w	w	u	v	b	a	e

This table shows that the six points on the circle form a *group* under these six operations.

¶ 5. THE SETS H, K OF CHAPTER V IDENTIFIED

The table obtained in ¶ 4 is identical (apart from an interchange of the names a, b) with that given on p. 61.

Further, the analysis of structure on p. 64 divided the group into the two cosets

$$H \equiv \{e, a, b\}, \quad \text{a subgroup,}$$

and $$K \equiv \{u, v, w\}.$$

This division receives pictorial emphasis here by the division into the two triangles EAB and UVW, each of which is, in fact, equilateral. The splitting of the set of six elements on p. 64 into two sets of three is the same as the division of the hexagon $EUAVBW$ into the two triangles EAB, UVW.

EXAMPLE

Prove that the analogous division into sets starting from the subgroups $\{e, u\}$, $\{e, v\}$, $\{e, w\}$ respectively gives precisely the three sets of parallel lines shown in the diagram (Fig. 4).

II. A GROUP OF EIGHT ELEMENTS

A closely analogous treatment may be applied to obtain a group of eight points on a circle. The notation is chosen so as to tempt the reader to obtain further generalisations for himself. (See the Examples at the end.)

¶ 6. We are now given a straight line u_1 and the resulting straight angle is divided into equal quarters by the lines u_2, u_3, u_4.

In the plane of the lines a circle is given and an arbitrary point E upon it. Points $U_1, A_2, U_2, A_3, U_3, A_4, U_4$ are defined exactly as before by means of lines parallel to $u_1, u_2, u_3, u_4, u_1, u_2, u_3$ in turn. The basic results for further work are set as examples for the reader to solve:

EXAMPLES

Prove that

    (i) $U_4 E \parallel u_4$;

    (ii) $U_4 A_2 \parallel A_4 U_2 \parallel u_1$,

           $U_4 A_3 \parallel E U_2 \parallel u_2$,

           $U_1 A_3 \parallel E U_3 \parallel u_3$,

           $U_1 A_4 \parallel E U_4 \parallel u_4$.

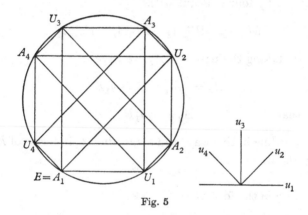

Fig. 5

In other words, *the sixteen lines in Fig. 5 fall into four sets of four each, in the directions of* u_1, u_2, u_3, u_4.

¶ 7. THE OPERATIONS AND THEIR PRODUCTS

By strict analogy with the proceeding case, we define four *operators* u_1, u_2, u_3, u_4 having, as before, the basic relations

$$u_1^2 = u_2^2 = u_3^2 = u_4^2 = e,$$

where e is the identity operator.

Note, in particular, that, in virtue of the parallelisms just found (Example (ii) above), the four points U_1, U_2, U_3, U_4 are obtained from E by the relations

$$U_1 = u_1 E, \quad U_3 = u_3 E,$$

$$U_2 = u_2 E, \quad U_4 = u_4 E.$$

The derivation of A_2, A_3, A_4 from E is less immediate, and products of operators are involved.

Consider, for example, the point A_2. It can be reached from E by four relevant paths:

$$EU_1 A_2, \quad EU_2 A_2, \quad EU_3 A_2, \quad EU_4 A_2.$$

Now, taking the path $EU_1 A_2$, we have

$$A_2 = u_2 U_1, \quad U_1 = u_1 E,$$

so that $$A_2 = u_2 u_1 E.$$

Proceeding in this way, and expressing A_2 in terms of E by the notation

$$A_2 = a_2 E,$$

we obtain the formulae

$$u_2 u_1 = u_3 u_2 = u_4 u_3 = u_1 u_4 = a_2;$$

and, similarly,

$$u_3 u_1 = u_4 u_2 = u_1 u_3 = u_2 u_4 = a_3,$$

$$u_4 u_1 = u_1 u_2 = u_2 u_3 = u_3 u_4 = a_4.$$

We therefore have a set of eight operators:

$$e, a_2, a_3, a_4, u_1, u_2, u_3, u_4.$$

EXAMPLES

1. Prove that the eight operators can all be expressed in terms of e, a_2, u_1 in the form

$$a_3 = a_2^2, \quad a_4 = a_2^3;$$
$$u_2 = a_2 u_1, \quad u_3 = a_2^2 u_1, \quad u_4 = a_2^3 u_1.$$

2. Establish the following 'multiplication table':

	e	a_2	a_3	a_4	u_1	u_2	u_3	u_4
e	e	a_2	a_3	a_4	u_1	u_2	u_3	u_4
a_2	a_2	a_3	a_4	e	u_2	u_3	u_4	u_1
a_3	a_3	a_4	e	a_2	u_3	u_4	u_1	u_2
a_4	a_4	e	a_2	a_3	u_4	u_1	u_2	u_3
u_1	u_1	u_4	u_3	u_2	e	a_4	a_3	a_2
u_2	u_2	u_1	u_4	u_3	a_2	e	a_4	a_3
u_3	u_3	u_2	u_1	u_4	a_3	a_2	e	a_4
u_4	u_4	u_3	u_2	u_1	a_4	a_3	a_2	e.

(The table establishes the eight operations as forming a group; it is known as the *dihedral* group.)

If, following the pattern of chapter v, we split the group into the two cosets from the subgroup $\{e, a_2, a_3, a_4\}$, we obtain the pair

$$H \equiv \{e, a_2, a_3, a_4\}$$

and
$$K \equiv \{u_1, u_2, u_3, u_4\}.$$

Now the product of an H by an H is always an H, of an H by a K is always a K, and so on. This leads to the table

	H	K
H	H	K
K	K	H

and the two subdivisions correspond pictorially to the squares $EA_2A_3A_4$ and $U_1U_2U_3U_4$ in Fig. 5.

This group is, in fact, identical in structure with that of eight elements given on p. 71, but the notation tends to cloud the issue. In order to make the correspondence clearer, we may re-write the present table in the form:

	e	u_2	u_4	a_3	u_1	a_2	a_4	u_3
e	e	u_2	u_4	a_3	u_1	a_2	a_4	u_3
u_2	u_2	e	a_3	u_4	a_2	u_1	u_3	a_4
u_4	a_4	a_3	e	u_2	a_4	u_3	u_1	a_2
a_3	a_3	u_4	u_2	e	u_3	a_4	a_2	u_1
u_1	u_1	a_4	a_2	u_3	e	u_4	u_2	a_3
a_2	a_2	u_3	u_1	a_4	u_2	a_3	e	u_4
a_4	a_4	u_1	u_3	a_2	u_4	e	a_3	u_2
u_3	u_3	a_2	a_4	u_1	a_3	u_2	u_4	e

The reconciliation with p. 71 is given by the relations

$$e = e, \quad a = u_2, \quad b = u_4, \quad c = a_3,$$
$$h = u_1, \quad u = a_2, \quad v = a_4, \quad w = u_3.$$

Corresponding to the cosets (p. 71)

$$H \equiv \{e, a, b, c\}, \quad K \equiv \{h, u, v, w\},$$

we now have

$$\{e, u_2, u_4, a_3\}, \quad \{u_1, a_2, a_4, u_3\},$$

giving the vertices

$$\{E, U_2, U_4, A_3\}, \quad \{U_1, A_2, A_4, U_3\}$$

of the rectangles $EU_2A_3U_4$ and $U_1A_2U_3A_4$.

We also obtained (p. 72) a subgroup $H \equiv \{e, c\}$, starting from which there were four cosets

$$H \equiv \{e, c\}, \quad K \equiv \{u, v\}, \quad L \equiv \{h, w\}, \quad M \equiv \{a, b\},$$

which formed a proper multiplication table. The cosets here are

$$H \equiv \{e, a_3\}, \quad K \equiv \{a_2, a_4\},$$
$$L = \{u_1, u_3\}, \quad M \equiv \{u_2, u_4\},$$

giving the four *diameters*

$$EA_3, \; A_2A_4, \; U_1U_3, \; U_2U_4$$

of the circle.

EXAMPLES

1. Starting from five lines u_1, u_2, u_3, u_4, u_5, obtain similarly a figure of ten vertices E, U_1, A_2, U_2, A_3, U_3, A_4, U_4, A_5, U_5 on a circle. Prove that, for example,

$$EU_1 \| A_2 U_5 \| A_3 U_4 \| A_4 U_3 \| A_5 U_2.$$

Taking the identity operator e as before, obtain a group by establishing the existence of the following relations:

$$u_2 u_1 = u_3 u_2 = u_4 u_3 = u_5 u_4 = u_1 u_5 = a_2,$$

$$u_3 u_1 = u_4 u_2 = u_5 u_3 = u_1 u_4 = u_2 u_5 = a_4,$$

$$u_4 u_1 = u_5 u_2 = u_1 u_3 = u_2 u_4 = u_3 u_5 = a_3,$$

$$u_5 u_1 = u_1 u_2 = u_2 u_3 = u_3 u_4 = u_4 u_5 = a_5.$$

Give the group table, and examine cosets.

2. Attempt to generalise these results to groups of $2n$ elements.

3. $A_1 A_2 A_3 A_4 A_5$ is a regular plane pentagon. Rotation through an angle $\frac{2}{5}\pi$ in its plane about its centre is an operation denoted by the symbol ω. Verify that the five operations denoted (in obvious notation) by e, ω, ω^2, ω^3, ω^4 all leave the pentagon as a whole unaltered, though vertices are interchanged. Prove that $\omega^5 = e$, and establish that the five operations form a group.

A further operation α is defined by 'turning the pentagon over' about the diameter through A_1. Prove that, once again, the pentagon as a whole is unaltered. Verify that the ten operations

$$e, \omega, \omega^2, \omega^3, \omega^4, \alpha, \omega\alpha, \omega^2\alpha, \omega^3\alpha, \omega^4\alpha$$

form a group essentially the same as that obtained in Example 1.

4. Generalise this also to obtain groups of $2n$ elements.

REMARKS. The interpretation implied in Example 3 is the one usually adopted for this sequence of groups. There does, however, seem to be entertainment value in relating it also to the configurations obtained by argument from very elementary Euclidean geometry. The production of such instances from work that is essentially familiar is one of the main themes of this book.

CHAPTER VII

AN ABSTRACT STRUCTURE
FOR ANGLES

The purpose of this chapter is to familiarise the reader
with the idea that standard notation, and processes
associated with that notation, can be extended far
beyond its ordinary uses, provided that a clearly defined
meaning is provided at each stage. The actual mathematical knowledge required is very elementary.

⟦ 1. We are concerned with *lines* and the *angles* between
them. For reference, lines will be named in italics,

$$a, b, c, \ldots, u, v, \ldots$$

and so on.

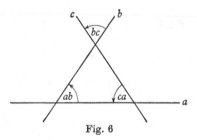

Fig. 6

⟦ 2. A symbol such as

$$ab$$

will denote the angle turned through *in the counterclockwise sense* by a line starting at position a and
reaching for the first time the position b.

The diagram (Fig. 6) shows in illustration the three
angles bc, ca, ab.

Each angle defined in this way ('reaching *for the first time* the position *b*') is necessarily less than 180°, or, in *radian measure* which we shall always adopt,

$$ab < \pi.$$

The angle between two *parallel* lines, including as a special case 'the angle between a line and itself' will be regarded as zero, since no turning is required. Thus

$$u \,\|\, v \quad \Leftrightarrow \quad uv = 0;$$

and, always, $\qquad uu = 0.$

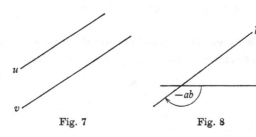

Fig. 7   Fig. 8

The symbol $\qquad -ab$

is used naturally for the angle from *a* to *b* in the *clockwise* sense. It is then automatic that

$$-ab = ba.$$

Thus *the order in which the letters are written is very important.*

¶ 3. The *sum* $\qquad ab + bc$

of two angles *ab*, *bc* with a common 'arm' *b* is in fact equal to the angle $\qquad ac,$

since it is the turning from a to b followed by the turning from b on to c. Thus we have the algebraic theorem

$$ab + bc = ac.$$

For lines positioned as in Fig. 9, this result is merely a statement in algebraic form of the fact that the exterior angle of a triangle is equal to the sum of the two interior opposite angles—but *counterclockwise sense is now vital*.

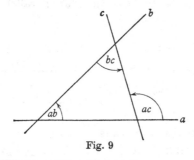

Fig. 9

By making c coincide with a, as it may, we have the result

$$ab + ba = 0.$$

This result, taken with the formula

$$-ab = ba$$

at the end of ¶ 2, shows that angles may be manipulated for summation by following the rules of ordinary algebra. Thus

$$ab + ba = 0 \Leftrightarrow ab = -ba.$$

Simple extension of the preceding work yields the formulae

$$ab + bc + cd + \ldots + uv = av,$$

$$ab + bc + cd + \ldots + ua = 0,$$

governing larger numbers of angles.

Note, in particular, that, *if a, b, c are the sides of a triangle, then* $ab + bc + ca = 0;$

that is, the sum of the angles is *zero* when all are taken in the counterclockwise sense of the symbols ab, bc, ca.

¶ 4. PARALLEL LINES

Suppose that u, v are two given parallel lines and that a is a transversal. Since the lines are parallel, we have

$$uv = 0$$

$$\Rightarrow ua + av = 0$$

$$\Rightarrow ua = -av$$

$$\Rightarrow ua = va.$$

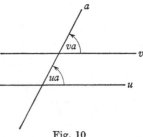

Fig. 10

In Fig. 10, the two *corresponding angles* are thus equal—a well-known property.

Suppose, conversely, that three lines u, v, a are so related that $ua = va.$

Then $ua = -av$

$$\Rightarrow ua + av = 0$$

$$\Rightarrow uv = 0$$

$$\Rightarrow u \parallel v.$$

Hence *a necessary and sufficient condition for two lines u, v to be parallel is that, if a is any transversal cutting them, then* $ua = va,$

the alternative form $au = av$

being equally permissible.

This suggests a convention which can often be useful: *if u, v are two parallel lines, we agree to write*

$$u = v.$$

(Waverers may interpret this by saying that the *direction* of u is equal to the *direction* of v.)

We then have the algebraic sequences:

$$u = v$$
$$\Rightarrow ua = va, \quad au = av$$
and
$$au = av \quad \text{or} \quad ua = va$$
$$\Rightarrow \quad u = v.$$

It is almost as if the symbol a were to 'cancel' as in elementary algebra.

Note that, if u, v, w are three given lines such that u, v are parallel and u, w are parallel, then

$$u = v$$
and
$$u = w.$$

But it is then known, as a geometrical theorem, that v and w are parallel, so that

$$v = w.$$

That is, we are entitled to use the algebraic sequence

$$u = v, \quad u = w$$
$$\Rightarrow v = w.$$

⟦ 5. THE RIGHT-ANGLE COMPLICATION

An unexpected difficulty occurs in the study of right angles. Suppose that a, b are two given lines such that

$$a \perp b.$$

Then the two angles, ab, ba, measured in the counter-clockwise sense, are equal, so that

$$ab = ba.$$

Thus, *if two lines a, b are perpendicular, then*

$$ab = ba.$$

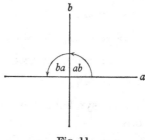

Fig. 11

On the other hand, we have the general result

$$ab = -ba,$$

so that, for right angles, we have the two relations

$$ab + ba = 0,$$

$$ab - ba = 0,$$

but
$$ab \neq 0.$$

At first sight this seems strange, but the phenomenon is closely akin to the *divisors of zero* which we studied in chapter I. In fact,

$$\begin{cases} ab + ba = 0, \\ ab - ba = 0 \end{cases}$$

$$\Rightarrow 2ab = 0$$

but *it is possible for 2ab to be zero when ab is not itself zero.*

In order to give a numerical analogy, suppose that p, q are any two numbers of ordinary arithmetic, and use the symbol

$$p \times q$$

to denote the remainder after dividing the ordinary product pq by 30; thus

$$7 \times 9 = 3,$$

$$8 \times 7 = 26.$$

Consider, in particular, the two numbers 3 and 5. In this arithmetic, we have

$$(3 \times 5) + (5 \times 3) = 0,$$

and $\qquad (3 \times 5) - (5 \times 3) = 0$

but, nevertheless, $\qquad 3 \times 5 \neq 0.$

We may, indeed, add the two equations to give the relation

$$2(3 \times 5) = 0,$$

which is *true*; but the result $3 \times 5 = 0$ *does not follow*. In fact, 15 is a *divisor of zero* in this arithmetic, since

$$2 \times 15 = 0,$$

$$15 \neq 0.$$

The analogy between this arithmetic and the 'arithmetic of angles' is perhaps worth a brief comment. In the former, a number such as 283 is reduced by as many multiples of 30 as possible, to give 13; in the latter, an angle such as $\frac{10}{3}\pi$ (obtained, say, by additions) is reduced by as many multiples of π as possible, to give $\frac{1}{3}\pi$. An equation like

$$2x = 0$$

in the former not only has the zero solution, but also the solution $x = \frac{1}{2} \cdot 30 = 15$; an equation like

$$2x = 0$$

in the latter not only has the zero solution, but also the solution $x = \frac{1}{2}\pi$.

¶ 6. CYCLIC QUADRILATERALS

Suppose that four points P, U, L, X lie on a circle (Fig. 12). They can be joined by six lines

$$UL, LX, XU, PX, PU, PL$$

which we may call x, u, l, y, v, m

respectively. (The names are deliberately chosen to be unsystematic in the exposition; in applications a more symmetrical notation would naturally be adopted, as in the Illustrations which follow.)

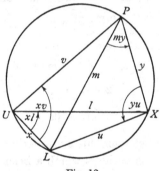

Fig. 12

Now select the four points in any order, for example, L, P, X, U; and write down the four sides which 'follow that order of the letters',

$$LP, PX, XU, UL,$$

returning at the end to the first letter to complete the cycle. These lines, in order, are

$$m, y, l, x.$$

Finally write down the product, still in order,

$$mylx,$$

put a plus sign in the middle, and equate to zero:

$$my + lx = 0.$$

Written in the alternative form

$$my = xl,$$

this is the familiar theorem, *angles in the same segment are equal*, as a glance at the diagram verifies.

Follow now the identical process for the alternative ordering of vertices P, X, L, U:

$$PX, XL, LU, UP,$$

$$yuxv,$$

$$yu + xv = 0.$$

This is the theorem: *the opposite angles of a cyclic quadrilateral are supplementary*, as, once again, the diagram verifies.

It can be verified further that *all choices of order lead to one or other of these two theorems*, which are therefore essentially equivalent.

EXAMPLES

Verify the result for each of the orders:

(i) U, L, X, P.   (ii) P, U, X, L.   (iii) P, U, L, X.

Remembering that the *converses* of the two 'angles' theorems are also true, we can reverse the final step when necessary and deduce that *if, say,*

$$my + lx = 0$$

or $$yu + xv = 0,$$

then the corresponding quadrilateral is cyclic.

The power of this method may be illustrated by one or two riders:

(i) *Let ABC be a given triangle and P, Q, R points on the sides BC, CA, AB respectively. To prove that, if the circles PBR, PCQ meet in U, then AQUR is also cyclic.*

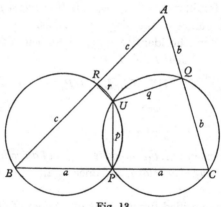

Fig. 13

Name the sides as follows:

$BC = a$, $CA = b$, $AB = c$; $UP = p$, $UQ = q$, $UR = r$.

(Of course BP and PC are each the line a.)

From the cyclic quadrilateral $PUQC$, we choose the sequence $$P, U, Q, C,$$

giving $PU, UQ, QC, CP,$

$$p, q, b, a,$$

$$pqba,$$

$$pq + ba = 0.$$

Similarly, from $PURB$, we have

$$pr + ca = 0,$$

so that, reversing the sense,

Hence, by addition, $rp + ac = 0.$

$$rp + pq + ba + ac = 0,$$

so that $rq + bc = 0,$

and so the quadrilateral $RUQA$ is cyclic, as required.

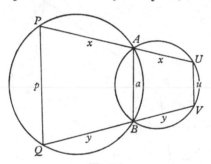

Fig. 14

(ii) *Two circles cut in A, B. A line through A cuts the first in P and the second in U; a line through B cuts the first in Q and the second in V. To prove that PQ is parallel to UV.*

Expressed with more normal brevity, the argument is: With the notation of Fig. 14,

$$BAPQ \text{ cyclic} \quad \Rightarrow \quad ax + py = 0,$$
$$BAUV \text{ cyclic} \quad \Rightarrow \quad ax + uy = 0.$$

Hence $$py = uy,$$

so that (p. 97) $$p = u,$$
and so $PQ \parallel UV$.

(iii) THE EXTENSION OF SIMSON'S LINE
(A harder example.)

Let ABC be a given triangle and U a point on its circum-circle. Points P, Q, R are taken on BC, CA, AB respectively so that, for angles measured in the counterclockwise

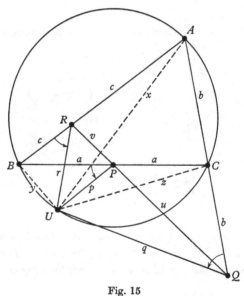

Fig. 15

sense, the angles between BC and UP; between CA and UQ; between AB and UR are equal. To prove that the points P, Q, R are collinear.

We use the notation

$$BC = a, \quad CA = b, \quad AB = c;$$
$$UP = p, \quad UQ = q, \quad UR = r;$$
$$UA = x, \quad UB = y, \quad UC = z;$$
$$PQ = u, \quad PR = v.$$

Then we are *given* that

$$ap = bq = cr.$$

Now
$$ap = bq$$
$$\Rightarrow ap + qb = 0$$
$$\Rightarrow CPUQ \text{ cyclic}$$
$$\Rightarrow QPUC \text{ cyclic}$$
$$\Rightarrow up + zb = 0;$$

and
$$ap = cr$$
$$\Rightarrow ap + rc = 0$$
$$\Rightarrow BPUR \text{ cyclic}$$
$$\Rightarrow RPUB \text{ cyclic}$$
$$\Rightarrow vp + yc = 0.$$

But
$$UCAB \text{ cyclic}$$
$$\Rightarrow zb + cy = 0.$$

Hence, from these equations,

$$up = bz$$
$$= cy$$
$$= vp.$$

Hence
$$u = v,$$

so that the lines u, v are parallel or coincident. But they have a point P in common, and so they are coincident.

Hence P, Q, R are collinear.

EXAMPLE

1. Points A, B, P, Q, L, M taken in that order round a circle are such that

$$AQ \parallel BP, \quad AM \parallel BL.$$

Prove that $$PM \parallel QL.$$

(One method considers the cyclic quadrilaterals $MPQA$, $PBLM$, $LQPM$, together with 'equalities' from the parallel lines.)

¶ 7. ISOSCELES TRIANGLES AND ANGLE BISECTORS

Let UA, UB be two given lines and UX *one of* the bisectors of the angles between them. Let UA, UB, UX be the lines a, b, x respectively. Then

$$ax = xb,$$

or $$ax + bx = 0.$$

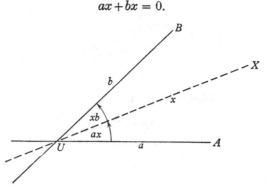

Fig. 16

Reversing the argument, we can prove that, if

$$ax = -bx,$$

then x is *one of* the bisectors of the angle between a and b.

Thus x is a bisector $\Rightarrow ax = -bx$;

$$ax = -bx \Rightarrow x \text{ is } one\ of \text{ the bisectors.}$$

This algebra does not distinguish between the two bisectors.

Suppose now that ABC is an *isosceles triangle* in which

$$AB = AC.$$

Write

$$BC = a, \quad AB = b, \quad AC = c.$$

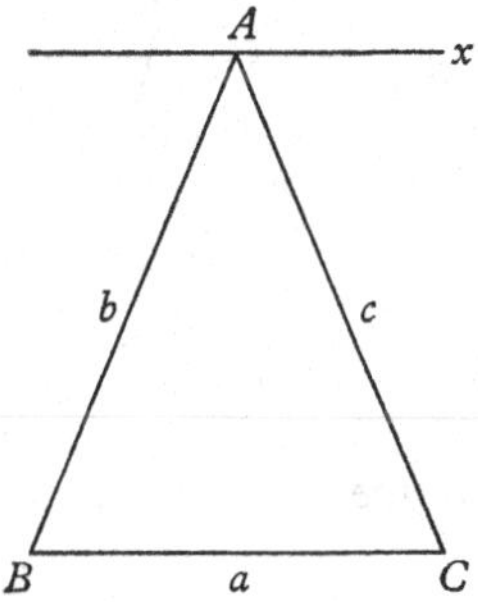

Draw the line x through A parallel to BC, so that (p. 97)

$$x = a.$$

Since the triangle is isosceles, x is a bisector (the external one in Fig. 17) of the angle between b and c; hence

$$xb + xc = 0.$$

Fig. 17

But $x = a$, and so *the 'base' angles satisfy the relation*

$$ab + ac = 0.$$

The *converse*, that

$$ab + ac = 0$$

$\Rightarrow$ triangle ABC isosceles with $AB = AC$

follows by similar argument.

For example:

Given that AB, PQ are two parallel chords of a circle, to prove that $AP = BQ$.

Draw AL parallel to BQ meeting PQ in L. Write

$$AB = a, \quad PQ = p, \quad AP = x, \quad BQ = y, \quad AL = u.$$

Then
$$AB \parallel PQ \quad \Rightarrow \quad a = p,$$
$$ABQP \text{ cyclic} \quad \Rightarrow \quad ay + px = 0,$$
$$AL \parallel BQ \quad \Rightarrow \quad u = y.$$

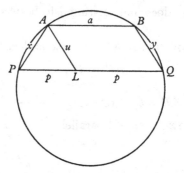

Fig. 18

Hence
$$ay + px = 0$$
$$\Rightarrow py + px = 0$$
$$\Rightarrow pu + px = 0$$
$$\Rightarrow AP = AL$$
$$= BQ.$$

CHAPTER VIII

A VERY ABSTRACT ALGEBRA FOR ANGLES

While it is hoped that the account to be given in this chapter will be self-contained, so that no previous knowledge is essential, it will probably be true that the contents will give most pleasure—and bewilderment—to a reader who has a little knowledge of determinants and vector products; only definitions and very first principles are used, and these are, in any case, given in the text.

¶1. Determinants of Order Two

The symbol

$$\begin{vmatrix} a & b \\ c & d \end{vmatrix}$$

is often used to denote the expression

$$ad - bc,$$

known as a *determinant*. As an illustration of how it can arise, note that, if the two equations

$$ax + b = 0$$

and

$$cx + d = 0$$

have a common value, then (assuming that a and c are not zero) that value is given equally by

$$-b/a \quad \text{and} \quad -d/c,$$

so that

$$-b/a = -d/c,$$

or

$$ad - bc = 0.$$

We require one minor complication not used in normal determinant theory: we always name a product so that the first symbol in it is taken from the *upper* row. Thus

$$\begin{vmatrix} u & v \\ x & y \end{vmatrix}$$

$$= uy - vx,$$

with u before y and v before x.

¶ 2. VECTORS AND VECTOR PRODUCTS

All we require of a vector is that it shall be a set of numbers selected in a definite order. If, for instance, those numbers are a, b, c, then the vector is simply the triplet of numbers written in the notation

$$(a, b, c).$$

The vectors (a, b, c), (b, a, c) are quite different—though there is no reason why, in a special case, a and b should not be the same.

In normal vector theory, a, b, c are usually magnitudes, such as components of velocity. The manipulation of vectors is subject to closely defined rules. For us, however, a, b, c will not appear as magnitudes, and, fortunately, the normal rules of manipulation will not be required.

What we do require, however, is a function that usually arises fairly late in vector theory. Suppose that (a, b, c) and (p, q, r) are two given vectors. Their *vector product* is defined to be the vector

$$(br - cq, \; cp - ar, \; aq - bp).$$

The three elements are

$$br - cq \equiv \begin{vmatrix} b & c \\ q & r \end{vmatrix},$$

$$-(ar - cp) \equiv -\begin{vmatrix} a & c \\ p & r \end{vmatrix},$$

$$aq - bp \equiv \begin{vmatrix} a & b \\ p & q \end{vmatrix},$$

where the three determinants arise from the array

$$\begin{array}{ccc} a & b & c \\ p & q & r \end{array}$$

by omitting successively the first, second and third columns. The negative sign at the second element is unfortunate but essential.

As an illustration of how these three determinants come together in normal practice, consider the two equations

$$ax + by + cz = 0,$$

$$px + qy + rz = 0.$$

It is easy to verify that the *ratios* of the three variables x, y, z satisfy the relations

$$\frac{x}{br - cq} = \frac{y}{cp - ar} = \frac{z}{aq - bp}.$$

As in ¶ 1, we always put the elements of the first vector before those of the second in each product; thus we write br and *not rb*.

¶ 3. APPLICATIONS IN GEOMETRY

The argument now becomes very abstract, though essentially simple.

Let A, B, C, D be four given coplanar points and use the notation

$$BC = a, \quad CA = b, \quad AB = c;$$
$$DA = p, \quad DB = q, \quad DC = r.$$

Form the 'vector product'

$$(br - cq, \quad cp - ar, \quad aq - bp),$$

and consider the first element

$$br - cq.$$

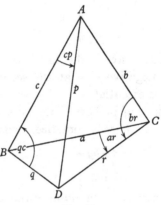

Fig. 19

For convenience, write it in the form

$$br + qc.$$

This is the *sum* of the angles at B and C, namely

$$\angle ACD + \angle ABD$$

when each is measured in the counterclockwise sense.

This property immediately directs our attention to the case when the quadrilateral $ABCD$ is *cyclic*; for then the sum of the angles is π.

Thus *a necessary and a sufficient condition for the quadrilateral ABCD to be cyclic is that the first component*

$$br - cq$$

is zero.

But now things move quickly. For the second component

$$cp - ar$$

is the difference between the 'same segment' angles

$$\angle BAD, \quad \angle BCD$$

and *it is a necessary and sufficient condition for the quadrilateral ABCD to be cyclic that these two angles are equal*: that is, *that*

$$cp - ar = 0.$$

In the same way *it is a necessary and a sufficient condition for the quadrilateral ABCD to be cyclic that*

$$aq - bp = 0.$$

Collating these three facts, we have the surprising result that *a necessary and a sufficient condition for the quadrilateral ABCD to be cyclic is that any one of the components of the 'vector product' shall be zero; and, consequently, when any one of the components is zero, so are all three.*

Note one point that we passed over without emphasis when establishing the notation: the 'vector' (a, b, c) has as its components the three sides of the triangle formed by three of the sides of the quadrilateral, and then the 'vector' (p, q, r) has as its components the three lines joining the *respective* vertices of that triangle to the fourth vertex of the quadrilateral.

¶ 4. THE ALTITUDES OF A TRIANGLE

Let ABC be a given triangle and draw the altitudes AP, BQ, CR perpendicular to the opposite sides BC, CA, AB. It is known that the lines AP, BQ, CR meet in a point H called the *orthocentre* of the triangle ABC. For notation, we write

$$BC = a, \quad CA = b, \quad AB = c;$$
$$AP = u, \quad BQ = v, \quad CR = w;$$
$$QR = p, \quad RP = q, \quad PQ = r.$$

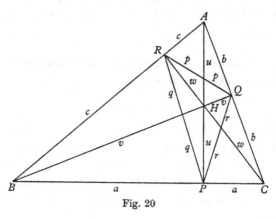

Fig. 20

By the right-angle properties, we have

$$au = ua = bv = vb = cw = wc.$$

Consider first the quadrilateral $AQHR$. Take HQR as basic triangle and A as fourth vertex. The 'vectors' are

$$(p, w, v), \quad (u, b, c)$$

and the 'vector product' is

$$(wc - vb, \; vu - pc, \; pb - wu).$$

By the right-angle relations, the first component $wc - vb$ is zero and so all components are zero. We may therefore write, symbolically,

$$(wc - vb,\ vu - pc,\ pb - wu) = 0.$$

Identical argument from the quadrilaterals $BRHP$, $CPHQ$ gives

$$(ua - wc,\ wv - qa,\ qc - uv) = 0,$$

$$(vb - ua,\ uw - rb,\ ra - vw) = 0.$$

In particular, we have

$$wv - qa = 0,\quad ra - vw = 0,$$

so that $\qquad\qquad vw = aq,\quad vw = ra.$

Thus $\qquad\qquad\qquad aq = ra,$

so that a is one of the bisectors of the angle between q and r; it follows that u, being perpendicular to a, is the other. Hence *BC and AP are the bisectors of the angle QPR.*

NOTE. This method does not of itself decide which is the internal and which the external bisector. The reader should draw an alternative version of Fig. 20, taking the angle ABC to be obtuse, and then compare results.

EXAMPLES

1. Repeat the preceding argument for the case when the angle ABC is obtuse.

2. Prove that H, in the given diagram, is the in-centre of the triangle PQR; and determine what happens when the angle ABC is obtuse.

¶ 5. THE SIMSON LINE AGAIN

Let ABC be a given triangle and U a point on its circumcircle. Draw UP, UQ, UR to be *perpendicular* (this time; compare p. 104) to BC, CA, AB respectively. It is required to prove that *the points P, Q, R are collinear.*

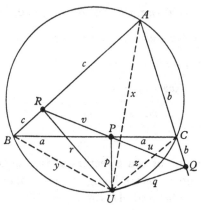

Fig. 21

Name the lines

$$BC = a, \quad CA = b, \quad AB = c,$$
$$UP = p, \quad UQ = q, \quad UR = r,$$
$$UA = x, \quad UB = y, \quad UC = z,$$
$$PQ = u, \quad PR = v.$$

The right-angles give the relations

$$ap = pa = bq = qb = cr = rc.$$

Consider first the quadrilateral $UABC$. Take ABC as basic triangle and U as fourth point. The vectors then are

$$(a, b, c), \quad (x, y, z)$$

so that, since the quadrilateral is given to be cyclic,

$$(bz - cy, \; cx - az, \; ay - bx) = 0.$$

Take next the quadrilateral $RPBU$, taking RPB as basic triangle and U as fourth point. The vectors are

$$(a, c, v), \quad (r, p, y).$$

The vector product is

$$(cy - vp, \; vr - ay, \; ap - cr).$$

Now $ap - cr$ is zero by the right-angles, and so

$$(cy - vp, \; vr - ay, \; ap - cr) = 0.$$

Similar argument from the quadrilateral $QPCU$ gives the vectors
$$(a, b, u), \quad (q, p, z),$$

so that $\quad (bz - up, \; uq - az, \; ap - bq) = 0.$

In particular, these zero vector products give respectively

$$bz - cy = 0, \quad cy - vp = 0, \quad bz - up = 0,$$

so that $\qquad up = bz = cy = vp.$

Hence (p. 97) $\qquad u = v,$

so that the lines u, v are parallel or coincident. But they have the point P in common, so that they coincide; that is, P, Q, R are collinear.

CHAPTER IX

SOME METRICAL ANALOGIES

¶ 1. The present chapter is probably the hardest in the book. It seems likely that the reader, if he will, may be able to modify it towards a simpler treatment; the effort would certainly be entertaining.

A new term may now be introduced to give precision to an idea that has been running through several parts of the work. Two integers a and b are said to be *equal, modulo n*, when they differ by an *exact integral multiple* of n. The notation used is

$$a \equiv b (\operatorname{mod} n),$$

meaning that an integer p can be found such that

$$a - b = pn.$$

This was the essential feature of digital arithmetic: instead of, say, a sum expressed in the form

$$7 + 9 = 16$$

it was, in effect, noted that

$$16 \equiv 6 \quad (\operatorname{mod} 10),$$

so that

$$7 + 9 \equiv 6 \quad (\operatorname{mod} 10).$$

The parenthesis was omitted, but was there 'in the spirit'.

In a similar way, the 'modulus' idea was behind the decision to reduce angles by multiples of π, so that, for instance, a statement such as

$$ab = cd$$

need not mean that the two angles ab, cd are equal (in the clockwise sense) but, more generally, that the difference

$$ab - cd$$

is an exact multiple of π. It is sometimes convenient to write

$$ab \equiv cd \quad (\bmod \pi)$$

to denote this relationship.

Much of the work that follows is simple, but it suggests ideas that become complicated. The difficulty arises from two features which will now be mentioned in turn.

¶ 2. A THEOREM AND ITS CONVERSE

The reader will certainly know about the distinction that must be drawn between a statement and its converse; but experience indicates that the distinction, though known in theory, is often ignored in practice, with results that are sometimes disastrous. Thus the statement, 'A multiple of 4 is a multiple of 2' is clearly true; but the *converse*, 'A multiple of 2 is a multiple of 4' is not true. In notation introduced earlier

n is a multiple of $4 \Rightarrow n$ is a multiple of 2;

but the sense of the arrow cannot be reversed.

Much of the discussion which follows in this chapter will move in one direction only, with consequences that sometimes seem curious.

¶ 3. DIAGRAMS TO ILLUSTRATE 'MODULAR ARITHMETIC'

The diagram (Fig. 22) illustrates clearly the familiar arithmetical fact that

$$6 + 8 = 14.$$

There seems, however, to be some difficulty in obtaining a diagram for the corresponding result in digital arithmetic,

$$6 + 8 = 4.$$

It can, of course, be done, but the complications appear to cancel the benefits.

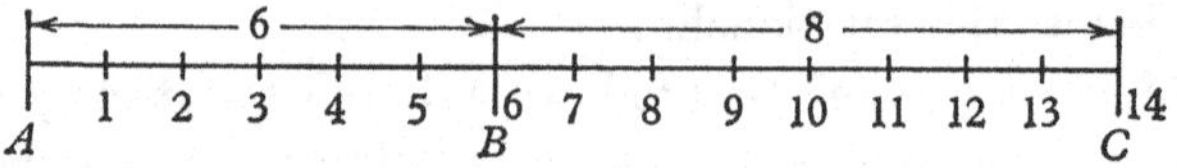

Fig. 22

A similar difficulty occurs in dealing with angles. If, say, ab and cd are two angles such that

$$ab = \tfrac{2}{3}\pi, \quad cd = \tfrac{3}{4}\pi,$$

then $\qquad ab + cd = \tfrac{5}{12}\pi \pmod{\pi}.$

It will be seen that we should like to be able to illustrate this relation by taking, somehow, $\tfrac{2}{3}\pi, \tfrac{3}{4}\pi, \tfrac{5}{12}\pi$ as segments of a *straight line* in such a way that the representation is significant. But this seems hard to achieve.

With these words of warning (which the reader will be wise to forget for the present), we proceed to the main theme of this chapter.

¶ 4. The Basic Analogy

The notation

$$ab$$

has been used to represent the angle *from* a line *a to* a line *b*, measured in the counterclockwise sense. The analogous notation

$$AB$$

represents the distance *from* a point *A to* a point *B*

measured in an agreed sense along the line. Just as the symbol

$$ba$$

represents a reversal of sense, giving the angle from b to a, so the symbol

$$BA$$

represents a reversal, giving the distance from B to A. The two relations

$$ab + ba = 0, \quad AB + BA = 0$$

are expressions of these facts.

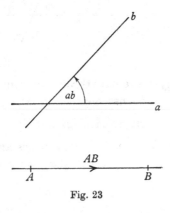

Fig. 23

In order to use these ideas, we shall suppose that some configuration of lines $a, b, c, \ldots$ is given, and we shall study *analogous* properties, as we shall call them, of systems of points $A, B, C, \ldots$, where the rules governing the relative positions of $A, B, C, \ldots$ will be defined by analogy with corresponding rules for $a, b, c, \ldots$.

¶ 5. THE TRIANGLE

Suppose that a, b, c are three given lines. The counter-clockwise angles bc, ca, ab are, for all positions of the lines, subject to the relation

$$bc + ca + ab = 0.$$

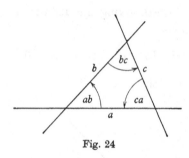

Fig. 24

For the analogy, we require three points whose distances apart, in an agreed sense, we wish to satisfy the relation

$$BC + CA + AB = 0.$$

At this point it seems to become necessary to adopt a convention, that, *to compare the magnitudes of two sensed*

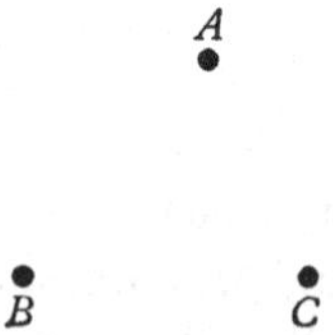

Fig. 25

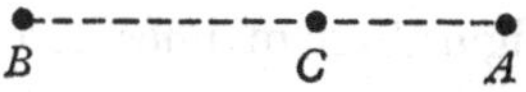

Fig. 26

straight lines (whether for equality or for purposes of addition or subtraction) *then those lines must be taken as coincident or parallel.* Equality on non-parallel lines will not be regarded as meaningful. Since, here, we wish to add BC to CA, we must take those lines as *coincident* (Fig. 26, and not Fig. 25). It is then evident that, *for all relative positions of A, B, C on that line, there is an identity*

$$BC + CA + AB = 0$$

when regard is had to sense. (Briefly, this relation is equivalent to the sequence of instructions: 'Go from B to C, then from C to A, then from A to B'. The total effect is zero.)

There is thus an automatic analogy between the relation

$$bc + ca + ab = 0$$

for angles in the counterclockwise sense and the relation

$$BC + CA + AB = 0$$

for distances between collinear points having regard to sense along the line. From the relation $bc + ca + ab = 0$ for angles in a given configuration we can deduce the relation $BC + CA + AB = 0$ for lines in the analogous configuration, *together with any consequences that follow from that deduction.*

In dealing with three given lines a, b, c only, the natural analogue has been three *collinear* points A, B, C. In more elaborate diagrams, line segments may, if desired, be compared on lines which are *parallel* rather than coincident—as in ⟦ 7.

⟦ 6. THE QUADRILATERAL

Suppose that a given quadrilateral has sides a, b, c, d. The analogue is a configuration of four points A, B, C, D.

(Strictly speaking, a figure defined by four points is called a *quadrangle* and a figure defined by four *lines* a *quadrilateral*; but the distinction is often blurred in work of this kind.)

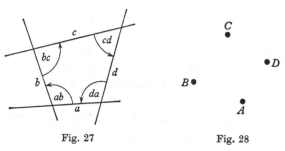

Fig. 27   Fig. 28

If we want to deal with any arbitrary selection of the possible angles between the given lines, then we shall have to select A, B, C, D to be collinear. But, as it happens, the cases that are most interesting for our purposes allow alternative selections, to which we now proceed.

¶ 7. THE CYCLIC QUADRILATERAL

Suppose next that the four lines a, b, c, d are the sides of a *cyclic* quadrilateral. The relations of interest to us were obtained previously (p. 100) and can be taken in the form

$$ab + cd = 0, \quad bc + da = 0$$

or, equivalently, $ab = dc, \quad bc = ad.$

We should therefore like to place the points A, B, C, D in such a way that

$$AB = DC, \quad BC = AD,$$

where we may allow the lines AB and DC to be parallel

rather than coincident and the lines BC and AD to be parallel rather than coincident if that proves more convenient. The attempt to do this leads at once to a *parallelogram ABCD*, so that *the analogue of a cyclic quadrilateral abcd may be taken to be a parallelogram ABCD.*

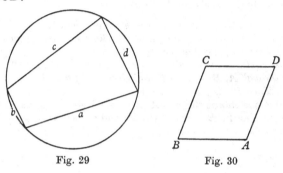

Fig. 29   Fig. 30

ILLUSTRATION. We return to a result used earlier (p. 102). *The lines a, b, c form a triangle ABC.* (To avoid confusion with the analogous figure, the names of points are not marked in Fig. 31.) *Points P, Q, R are taken on BC, CA, AB respectively,*

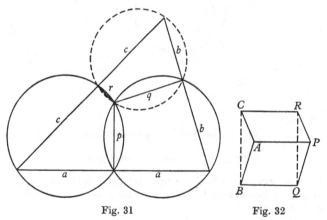

Fig. 31   Fig. 32

and the circles PBR, PCQ meet in U. Then the circle AQR also passes through U.

Denote UP, UQ, UR by the letters p, q, r.

The analogue of the cyclic quadrilateral *aprc* is a parallelogram *APRC*; the analogue of the cyclic quadrilateral *apqb* is a parallelogram *APQB*. Then *the analogue of the fact that bcrq is a cyclic quadrilateral is that BCRQ is a parallelogram*, a result that is easily verified otherwise.

HARDER ILLUSTRATION. An example which is easy to establish but whose significance is harder to grasp is given by another result proved earlier (p. 103).

Two circles meet in points A, B (not named in Fig. 33) and lines through A, B meet the first circle in P, Q and the second in U, V. Then PQ is parallel to UV.

Name the chords through A, B by the letters a, b; name the common chord AB by the letter x; and name PQ, UV by the letters p, u.

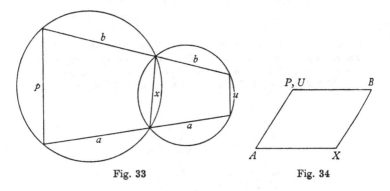

Fig. 33   Fig. 34

Then

$$axbp \text{ cyclic} \Rightarrow AXBP \text{ parallelogram},$$

$$axbu \text{ cyclic} \Rightarrow AXBU \text{ parallelogram}.$$

Hence *the two points P, U coincide*, the distance between them being zero by analogy with the fact that the angle between the lines p, u is zero.

The awkward feature is that the two cyclic quadrilaterals *axbp*, *axbu* are so related that they are represented by the *same* parallelogram.

The point of the apparent dilemma is that, if the lines *a*, *b*, *x* are given and are required to be the sides of a cyclic quadrilateral, then the *direction* of the fourth side is fixed—represented by the fourth vertex of the parallelogram—but its *position* is not. The lines *p*, *u* are two, among many, of such possible positions.

ILLUSTRATION. Let *a*, *b*, *c* be three lines forming a given triangle and *p*, *q*, *r* the lines joining the feet of the altitudes (Fig. 35). Then (by the converse of the 'angles in the same segment' theorem) the three quadrilaterals

are cyclic.   *abpc*,   *bcqa*,   *carb*

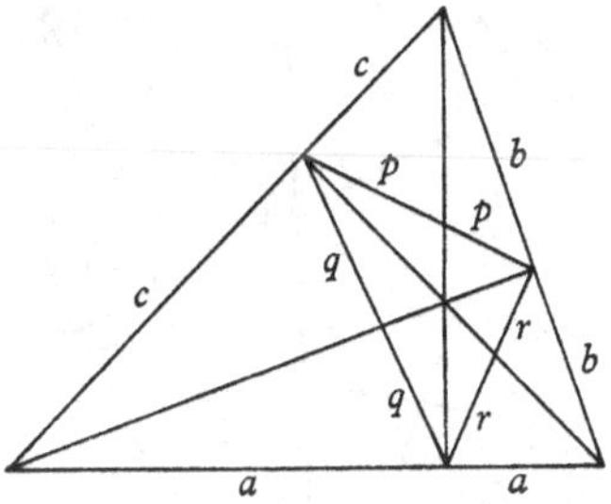

Fig. 35 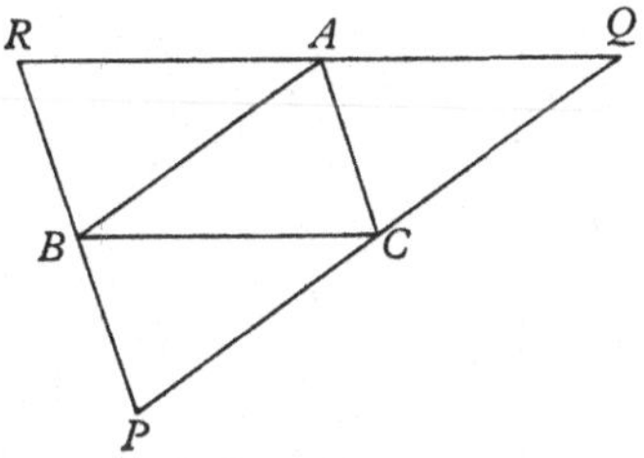   Fig. 36

The analogue is a triangle *ABC* with lines through the vertices parallel to the opposite sides forming a triangle *PQR*. Then the quadrilaterals

$$ABPC, \quad BCQA, \quad CARB$$

are parallelograms.

The two familiar figures of altitudes and middle points are therefore analogous.

EXAMPLE

Noting that the lines CB, QA, AR are equal, verify (by elementary geometry) the truth of a speculation that the angles cb, qa, ar are equal.

¶ 8. THE COMPLETE CYCLIC QUADRILATERAL

We have not yet given any attention to the analogue of the *diagonals* of a cyclic quadrilateral. For this we

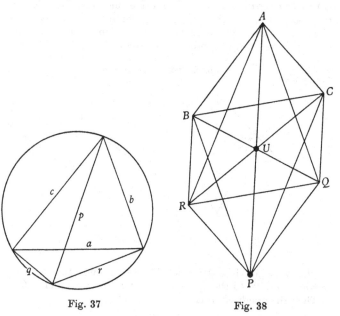

Fig. 37   Fig. 38

turn to the notation described in chapter VIII, ¶ 3 and now adopted in the diagram (Fig. 37). (Note that a, p are the diagonals and b, q and c, r the pairs of opposite sides.)

The six lines of the cyclic quadrilateral give us six points A, B, C, P, Q, R. The angle theorems for the circle give us

(i) two equalities based on the fact that the sum of the opposite angles is π, namely

$$bc = rq \quad \text{and} \quad br = cq;$$

(ii) four equalities based on the fact that angles in the same segment are equal, namely

$$ca = pr, \quad ab = qp, \quad cp = ar, \quad aq = bp.$$

Selecting these in convenient order, we have for the analogues:

$$BC = RQ, \quad BR = CQ,$$

so that *the four points B, C, Q, R are the vertices of a parallelogram*;

$$CA = PR, \quad CP = AR,$$

so that *the four points C, A, R, P are the vertices of a parallelogram*;

$$AB = QP, \quad AQ = BP,$$

so that *the four points A, B, P, Q are the vertices of a parallelogram.*

The result of this, bearing in mind that the diagonals of a parallelogram bisect each other, is that, *for the analogous figure derived from a cyclic quadrilateral, there exists a point U with respect to which the triangles ABC, PQR are mirror images*, in the sense that

$$AU = UP, \quad BU = UQ, \quad CU = UR.$$

(9. DIFFICULTIES

At the start of this chapter we gave warning of difficulties to come; they have now arrived.

In ¶ 8 we obtained a point U such that

$$AU = UP, \quad BU = UQ, \quad CU = UR.$$

This suggests that we might find a line u such that

$$au = up, \quad bu = uq, \quad cu = ur,$$

where, as usual, all lines parallel to u would do equally well for this purpose.

The theorem is, indeed, very nearly true—but *not quite*.

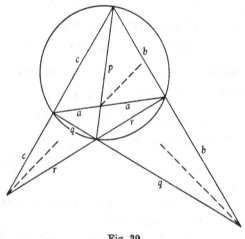

Fig. 39

The diagram (Fig. 39) repeats Fig. 37, where, however, the opposite sides b, q and c, r have been produced to meet. The relation

$$au = up$$

would require u as the direction of a bisector of the angle ap; the relation

$$bu = uq$$

would require u as the direction of a bisector of the angle bq; the relation

$$cu = ur$$

would require u as the direction of a bisector of the angle cr. The dotted lines indicate the directions of the 'obvious' bisectors; two of them seem parallel but the third does not. But the fact is that, indeed, *two of the bisectors are parallel and the third is perpendicular to them*.

EXAMPLE

Prove the statement just made, using the standard theorems of elementary geometry.

Once this result is established, we see at once that all that is necessary to produce a correct version of the theorem is to select the *alternative* bisector for the third angle. The dilemma is that the theory does not give us any method of making the selection from the 'analogue' diagram of parallelograms.

We must therefore give some attention to the problem of right angles and, as it will appear, of 'straight' angles also.

¶ 10. RIGHT ANGLES AND STRAIGHT ANGLES

The trouble about right angles lies embedded in the relation (see Fig. 40)

$$ab = ba$$

which, in essence, defines right angles, considered alongside the identity

$$ab + ba = 0,$$

which is a standard formula for all positions of b. We have, by addition,

$$2ab = 0 \quad \text{and} \quad 2ba = 0,$$

although neither ab nor ba is zero.

But *in the analogous diagram of points* (Fig. 41), *we cannot exhibit the relation*

$$AB = BA$$

except by taking each as zero. Thus *two perpendicular lines*

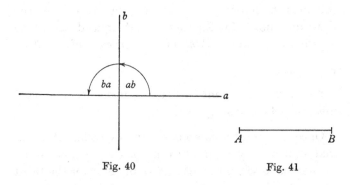

Fig. 40   Fig. 41

become *two coincident points*. This is the problem of 'one-way' argument to which we referred at the start:

$$a \perp b \quad \Rightarrow \quad A = B;$$

but, as we have often seen, *it is also true that*

$$a \| b \quad \Rightarrow \quad A = B.$$

Thus when we go *from* the figure of lines *to* the figure of points there is no ambiguity; we get, unexpectedly perhaps, coincident points. But in reasoning backwards we cannot make safe deductions without a check from the figure of lines itself—a given point may arise from any one of a set of parallel lines (that should not worry us greatly) *or*, equally, from any one of another set of parallel lines each member of which is perpendicular to each member of the first set.

To emphasise the point, consider the analogue of a *rectangle abcd*. Since

$$c \parallel a,$$

it follows that

$$C = A.$$

Since

$$b \perp a, \quad d \perp a,$$

it follows that

$$B = A, \quad D = A.$$

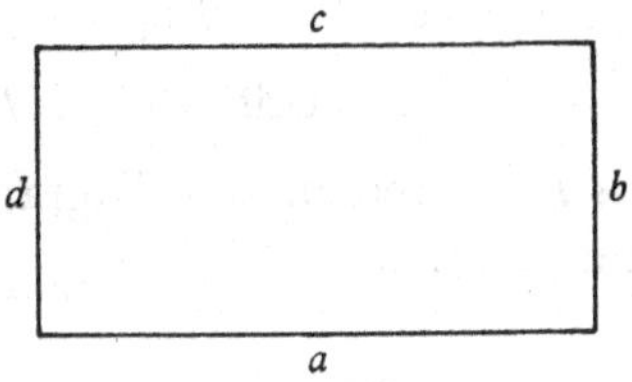

Fig. 42

Thus the four points A, B, C, D coincide.

In other words, the analogies must be handled with care. When this is done, some of the results are very striking, as the next two paragraphs demonstrate.

⟨ 11. Two Standard Results Related

We have already seen (p. 107) that if a, b, c are the sides of an isosceles triangle, with b, c along the equal sides, then there is a relation

$$ba - ac = 0$$

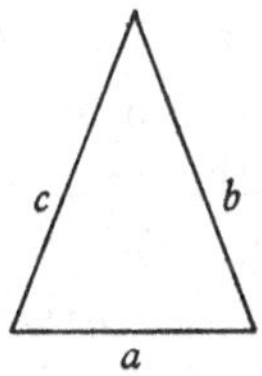

Fig. 43

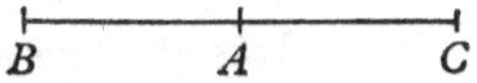

Fig. 44

which has for its analogue the equality

$$BA = AC,$$

so that *A is the middle point of BC.*

Suppose now that L, M are two points on a circle of centre O and that X is any point on the appropriate segment subtended by LM. Write

$$OL = u, \quad OM = v, \quad XL = a \quad XM = b, \quad OX = p,$$

as indicated in the diagram (Fig. 45). Then it is well known that

$$uv = 2ab.$$

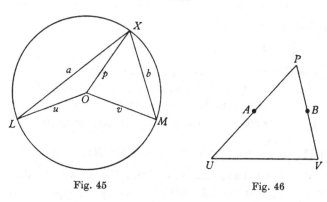

Fig. 45 Fig. 46

In the analogous figure, the fact that the triangle aup is isosceles gives A as the middle point of PU; and, similarly, B is the middle point of PV. Now UV is *parallel* to AB, so that (by the convention agreed on p. 122) the lengths of UV and AB are in the same proportions as the corresponding angles uv and ab. Hence *we have established the effective equivalence of the two theorems*:

(i) the angle subtended at the centre of a circle by an arc is double the angle at the circumference subtended by the same arc;

(ii) the line joining the middle points of the sides of a triangle is parallel to the base and equal to half the length of the base.

¶ 12. SIMSON'S LINE AGAIN

As a final example of the oddities that can occur with these analogies, consider once again the Simson line property.

From a point K on the circumcircle of a triangle ABC, perpendiculars KP, KQ, KR are drawn to the sides BC, CA, AB. Then the points P, Q, R are collinear (Fig. 47).

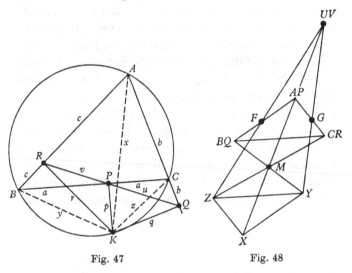

Fig. 47   Fig. 48

Name the lines as follows:

$$BC = a, \quad CA = b, \quad AB = c;$$
$$KP = p, \quad KQ = q, \quad KR = r;$$
$$KA = x, \quad KB = y, \quad KC = z;$$
$$PQ = u, \quad PR = v.$$

Since a, x are diagonals and b, y and c, z are pairs of opposite sides of a cyclic quadrilateral, the points $A, B, C,$

X, Y, Z in the analogous figure (Fig. 48) are so related (p. 128) that AX, BY, CZ have a common middle point M.

Since $p \perp a$, $q \perp b$, $r \perp c$, it follows that, in the analogous figure,

$$P \equiv A, \quad Q \equiv B, \quad R \equiv C.$$

(The argument may now be regarded as suspect, but seems to be correct.)

Since AQ and BP have a common middle point, say F, they arise from lines a, q and b, p which are opposite sides or diagonals in a cyclic quadrilateral; and inspection of the Simson line figure itself reveals that z, u is the third pair of lines of the quadrilateral, so that U is that point for which F is the middle point of UZ. In the same way, AR and CP have a common middle point G, leading to V as that point for which G is the middle point of VY. But it is easy to show that U and V coincide in a point such that A is the middle point of UM; and this coincidence of U and V, being the analogue of the identity of the lines u and v, is an expression of the property of the Simson line.

ANSWERS

PAGE 18. EXAMPLE 1.   (i) 1, 3;

(ii) 1, 4, 6, 9;

(iii) 2, 3, 7, 8;

(iv) 2, 4.

PAGE 34. EXAMPLES: 1, $\nleftrightarrow$;   2, $\Leftrightarrow$;   3, $\nleftrightarrow$;   4, $\nleftrightarrow$.

PAGE 39. EXAMPLE 2: Same as p. 37, with $e = 2$, $a = 4$, $b = 6$, $c = 8$.

PAGE 45. EXAMPLE 1: b, a, c.

EXAMPLE 2: b, a, u, v, w.

PAGE 47. EXAMPLE 2; $2^{-1} = 4$, $4^{-1} = 2$, $6^{-1} = 6$, $3^{-1} = 5$, $5^{-1} = 3$.

INDEX